Rajeev Ronanki, a problem solver and engineer, walks us through how AI will take our health into a *"Star Trek*-like future."* Twenty years ago, the *Star Trek* communicator went from a science fiction idea to a real device that fit in our pockets. Now, Rajeev and Anthem, Inc. want to put the entire Sickbay, *Star Trek's* medical center, into our pockets too.

A lover of puzzles and problem solving, Rajeev is tackling his biggest challenge yet—*the US healthcare system*—and he's armed with effective AI and backed by an innovation leader dedicated to improving health.

Just as Elon Musk used digital technologies to transform what was possible in aerospace and transportation, Rajeev is doing the same for human health and medicine.

Andrew Hessel
Founder, Humane Genomics, Inc.
Chairman, Genome Project-Write

If you've ever wondered what people mean by the "art of the possible" and how companies can turn innovation into business success, look no further. *You and AI* is the practical guide we all need to build the AI-first organization.

Erica Dhawan
Author, *Digital Body Language*

Rajeev had me at *You and AI*—a title that personifies the deeply personal, very human experience we are all collectively having with technologies that are changing and shaping our world, that are designed to improve our lives, our healthcare encounters, and our relationships with healthcare organizations.

In his conversational style, Rajeev explains why our understanding and relationship with AI as citizens and consumers is fundamental and critical as we consider and ensure our rights and responsibilities in a world heavily assisted and guided by AI, alongside its good friend blockchain. He also explains the inevitability of our reliance on technology and data to solve the growing complexity in healthcare and clearly defines that to achieve health outcomes that are improved, inclusive, and equitable it requires an equally hefty dose of proactive human care, participation, and intelligence.

Rajeev leads readers through the questions, complexities, breakthroughs, and concerns that prepare us as citizens to interact with AI and blockchain and puzzle together the future of healthcare. The title really says it all, and Rajeev's skilled storytelling is an enticing invitation to be one of the main characters!

Shawna Butler, RN, MBA
Nurse economist
Managing director, Exponential Medicine
Host, *See You Now* podcast

Rajeev Ronanki has deep experience in understanding, working with, and leveraging exponential technologies into impact. In this engaging read, he brilliantly sets the framework, illuminates the cutting edge, and elucidates the incredible potential for AI at its convergence with many other advancing fields to dramatically improve healthcare across the care continuum.

Daniel Kraft, MD
Founder and chair, Exponential Medicine

Ronanki lays out a clear, compelling, and realistic vision for how AI and blockchain can transform healthcare. He also describes what's already happening today with these technologies. There is no better guide to this complex set of behavioral, policy, and economic issues than *You and AI*.

Thomas H. Davenport
Distinguished professor, Babson College; visiting professor, Oxford University
Fellow, MIT Initiative on the Digital Economy
Senior advisor, Deloitte AI
Author, *The AI Advantage*; *Competing on Analytics*

RAJEEV RONANKI

YOU AND AI

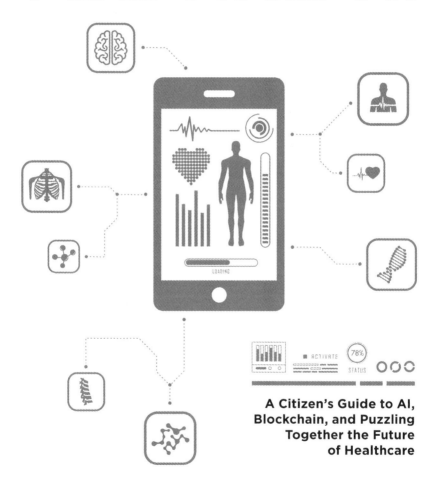

A Citizen's Guide to AI,
Blockchain, and Puzzling
Together the Future
of Healthcare

ForbesBooks

Published by ForbesBooks, Charleston, South Carolina.
Member of Advantage Media Group.

ForbesBooks is a registered trademark, and the ForbesBooks colophon is a trademark of Forbes Media, LLC.

Printed in the United States of America.

10 9 8 7 6 5 4 3 2 1

ISBN: 978-1-95086-342-6
LCCN: 2021910111

Cover design by David Taylor.
Layout design by Wesley Strickland.

This custom publication is intended to provide accurate information and the opinions of the author in regard to the subject matter covered. It is sold with the understanding that the publisher, Advantage|ForbesBooks, is not engaged in rendering legal, financial, or professional services of any kind. If legal advice or other expert assistance is required, the reader is advised to seek the services of a competent professional.

Advantage Media Group is proud to be a part of the Tree Neutral® program. Tree Neutral offsets the number of trees consumed in the production and printing of this book by taking proactive steps such as planting trees in direct proportion to the number of trees used to print books. To learn more about Tree Neutral, please visit **www.treeneutral.com**.

Since 1917, Forbes has remained steadfast in its mission to serve as the defining voice of entrepreneurial capitalism. ForbesBooks, launched in 2016 through a partnership with Advantage Media Group, furthers that aim by helping business and thought leaders bring their stories, passion, and knowledge to the forefront in custom books. Opinions expressed by ForbesBooks authors are their own. To be considered for publication, please visit **www.forbesbooks.com**.

For my mom and dad, and for Kelly, Jay, and the rest of the family.

Special thanks to Mariya Filipova.

CONTENTS

ACKNOWLEDGMENTS xi

FOREWORD 1
THE BEST WAY TO PREDICT THE FUTURE IS TO CREATE IT YOURSELF

CHAPTER 1 9
EXPONENTIAL TECHNOLOGIES AND OUR REVOLUTIONARY FUTURE

CHAPTER 2 23
THE BUSINESS OF EXPONENTIAL TECHNOLOGY

CHAPTER 3 43
THE HEALTHCARE PUZZLE: A SYSTEM UNDER STRAIN

CHAPTER 4 75
AI EXPLAINED

CHAPTER 5 97
ETHICS AND AI

CHAPTER 6 109

ARTIFICIAL INTELLIGENCE
AND HEALTHCARE

CHAPTER 7 131

BLOCKCHAIN EXPLAINED

CHAPTER 8 153

BLOCKCHAIN IN HEALTHCARE

CHAPTER 9 173

ALIGNING INCENTIVES ACROSS
THE HEALTHCARE INDUSTRY

CHAPTER 10 189

THE TECHNOLOGY ADOPTION CURVE

ACKNOWLEDGMENTS

THANKS TO THE publishing team—Laura Grinstead, Lauren Whitta-more, and Laura Rashley—and everyone that made this book possible.

Special thanks to Eric Barchard, Katy Feldner, Kristi Thomas, and George Atiyeh. Their tireless efforts made this book a much better product.

THE BEST WAY TO PREDICT THE FUTURE IS TO CREATE IT YOURSELF

I BELIEVE THAT a single person, driven by their massively trans-formative purpose (MTP), can change the world. We are living during a period of human history in which a scientist, leader, or entrepreneur—powered by their MTP and exponential technologies—can positively influence the lives of a billion people. Whether you are in financial services, real estate, insurance, healthcare, manufacturing, retail, advertising, or energy, understanding which exponential technologies are emerging and converging is critical to the longevity and effect of your business. This book offers a deep dive into the opportunity that the convergence of computation, sensors, networks, blockchain, and AI have to transform today's sick-care system into the promise of healthcare.

For most of human history, access to any reasonable healthcare was available only to the wealthiest among us. But technology is a force that transforms that which is scarce into something abundant. In much the same way that billions of people with a smartphone have free access to world information, in the decades ahead, AI, blockchain, sensors, and networks offer a future in which the best diagnostic healthcare is available to anyone, anywhere. This, then, is a vision for healthcare abundance.

In my recent book, *The Future Is Faster Than You Think*,[1] I outline several of the most powerful metatrends shaping the 2020s. Metatrends that have the ability to transform every industry on the planet in the next ten years. Metatrends that result from the convergence of exponential technologies, reinventing business models in their wake at a speed that will catch many of the traditional industry leaders by surprise. The following are the top five metatrends that I believe will reinvent the healthcare industry.

1. *Massive connectivity between all people and devices.* We are on the verge of connecting everyone and everything on the planet. In 2018, half of the world's population was plugged in online—3.8 billion out of today's 7.8 billion individuals. Over the decade ahead, the global deployment of 5G and satellite constellations, such as SpaceX's Starlink, will connect every human on Earth at multihundred-megabit speeds. In addition to connecting everyone, we are also connecting every digital smart device on the planet, something called the *Internet of Things (IOT)*. Today the IOT connects 35 billion devices, growing at a rate of 127 new devices per second, to 75 billion by 2025, onward to a trillion-dollar marketplace by 2027. Many of the devices being connected to the cloud and

---

1   Peter H. Diamandis, *The Future Is Faster Than You Think: How Converging Technologies Are Transforming Business, Industry, and Our Lives* (New York: Simon & Schuster, 2020).

to AI networks will be health related, allowing your physiology to be measured minute to minute throughout the year.

2. *Ubiquitous artificial intelligence embedded throughout our lives.* In the decade ahead, AI will be embedded into everything, everywhere, and will transform every business and industry. As noted by Sundar Pichai, CEO of Alphabet, "Artificial intelligence could have more profound implications for humanity than electricity or fire."[2] As reported by Gartner Predictions, AI will generate $2.9 trillion in business value and recover 6.2 billion hours of worker productivity in 2021.[3] In my humble opinion, there will be two kinds of companies at the end of this decade: those that are fully embracing and using AI and those that are out of business. As services like Alexa, Google Home, and Apple HomePod strengthen their AI and expand in functionality, such services will eventually become your round-the-clock diagnostic physician, monitoring the biomedical sensors in your bedroom, on your body and in your body, and your physiology to give you constant feedback and allow you to become the CEO of your own health.

3. *Significant increase in the human health span.* A dozen game-changing biotech and pharmaceutical solutions (currently in Phase 1, 2, or 3 clinical trials) will reach consumers this decade, adding at least an additional ten years to the human

---

2   Anthony Cuthbertson, "What's Bigger Than Fire and Electricity? Artificial Intelligence, Says Google Boss," *Newsweek*, April 5, 2021, https://www.newsweek.com/artificial-intelligence-more-profound-electricity-or-fire-says-google-boss-786531.

3   Katie Costello, "Gartner Says AI Augmentation Will Create $2.9 Trillion of Business Value in 2021," Gartner, August 5, 2019, https://www.gartner.com/en/newsroom/press-releases/2019-08-05-gartner-says-ai-augmentation-will-create-2point9-trillion-of-business-value-in-2021.

health span. Technologies include stem cell supply restoration, pathway manipulation, senolytic medicines, a new generation of endo-vaccines, peptide treatments, GDF11, supplementation of NMD/NAD+, among several others. And as machine learning continues to mature, AI is set to unleash countless new drug candidates, ready for clinical trials. This metatrend is driven by the convergence of genome sequencing, CRISPR technologies, gene therapies, AI, quantum computing, and cellular medicine. The end result, I believe, will be to make one hundred years old the new sixty, adding decades to our health span.

4. *Emergence of the spatial web.* One effect of the global COVID-19 pandemic has been the transition of the workplace from the physical office to the virtual experience. The same is happening to medical visits as we transition from healthcare in the hospital or doctor's office to healthcare remotely in the home. Much of this transformation is a result of virtual technologies such as Zoom, VR, AR, and embedded AI. In 2020, we saw the dramatic increase of Zoom's enterprise value to ten times the value of the top ten airlines *combined.* The year 2020 also saw a record investment of $2 billion in augmented and virtual reality technologies, with a projected ten times increase in AR and VR headsets between 2020 and 2024. The combination of VR/AR with artificial intelligence and blockchain will enable the emergence of the spatial web, or Web 3.0.

5. *Increasing abundance.* Although we are all bombarded with negative news from the media, on many fronts the world is getting better year after year with improvements such

as a reduction in global poverty, falling rates of childhood mortality, increasing access to electricity, and communications and internet services. Further, the convergence of high-bandwidth and low-cost communication and ubiquitous AI on the cloud is resulting in AI-aided education and AI-driven healthcare, which are being digitized and demonetized, becoming available to the rising billion on their mobile devices. Many of these positive changes are the result of exponential technologies transforming scarcity into abundance. As I remind the members of my Abundance 360 mastermind, the world's biggest problems are the world's biggest business opportunities. And fueling those entrepreneurs focusing on solving the world's grand challenges is an ever-increasing supply of capital abundance. As the *Economist* points out, companies raised more capital in 2020 (in the midst of a pandemic) *than at any time in human history.*[4] In 2020, US venture capitalist firms invested $156.2 billion in start-ups, equivalent to about $428 million every day of the year. This record sum was up from $136.5 billion invested in 2019. More bucks means more Buck Rogers, meaning that we will see more entrepreneurs taking more moonshots, starting more companies, and investing in more technologies.

So there you have it: just a few of the reasons why the decade ahead will look far different from the decade behind us. And why entrepreneurs and innovators armed with AI and blockchain, to name just a few of the technologies, have the potential to reinvent business models and industries like never before.

---

4   "Why COVID-19 Will Make Killing Zombie Firms Off Harder," *The Economist*, September 26, 2020, https://www.economist.com/finance-and-economics/2020/09/26/why-covid-19-will-make-killing-zombie-firms-off-harder.

Whether you are new to the AI, blockchain, and exponential technology universe, or you are an old hand looking for ground-breaking innovations, I am excited to introduce you to *You and AI*. This book will teach you about a future that is trending much more toward the medical world we envisioned during our youth watching *Star Trek* than anything around today. I invite you to join and explore what is unfolding at this very moment, a future that is faster—and better—than you ever thought possible.

—Peter H. Diamandis, MD

CHAPTER 1

EXPONENTIAL TECHNOLOGIES AND OUR REVOLUTIONARY FUTURE

AS A YOUNG CHILD, I always looked forward to the jigsaw puzzles my dad would bring home with him from work. My eyes would light up at the sight of the puzzle box tucked under his arm. I couldn't wait to dump the pieces out onto the table and splay them out with my fingers. Snapping the tiny cardboard shapes together was immensely satisfying. Each time the pieces fit, a rush of dopamine would shoot through my body, a tiny reward for a little problem solved.

I was methodical in my puzzling. Find the corner pieces first and put them into place. Separate the edges into their own pile and build the frame. Sort the rest by colors, like with like. Keep an eye out for logos and matching letters, any signage, the most obvious pieces first. Then test and retest until pieces start to fit. Work inward from the edge. Slowly, images emerge, order out of chaos. An elephant's tusk! The Eiffel Tower! The Taj Mahal's bulbous dome! The iconic circular hull of the *Starship Enterprise*! That one was my all-time favorite jigsaw puzzle. *Star Trek* was my favorite show at the time.

The family would all gather around the table to work on the puzzle together. But this was about more than just family time for me. My family would lose interest after a half hour or so and disperse once all the snacks had been eaten. My parents would wander off to watch television or read a magazine. Not me. I would sit up for hours finishing the puzzle. I could hardly wait for the next one. Sometimes I even undid the work so that we could do it over again.

As I got older, jigsaw puzzles lost their allure. They were too simple and no longer presented enough of a challenge. I lost interest in them. But I didn't lose interest in solving problems. I graduated from puzzles to building blocks. Legos opened up a whole new world of creative problem-solving. Jigsaw puzzles were a problem with one solution. One could arrive at that solution via various strategies, but the end point was preordained. Not so with Legos. I could build anything my mind could imagine. The sky was the limit.

My cousins treated the Legos more like toys. They would fashion little animals or dolls out of them to play make-believe with. I would play with them, but when they got bored, I would sit up and see what I could create with the blocks. The building itself was what kept me interested. They scratched that same itch the puzzles had.

My dad helped me use the Legos to build replicas of various things. We would tear pictures out of magazines and try to create the image in 3D. Famous buildings, household appliances, cars or rockets, little neighborhoods, and cityscapes. My crowning achievement was a miniature replica of the USS *Starship Enterprise*.

My dad, a mechanical engineer who worked on heavy machinery, was meticulous with the details. He helped me plan the parts and how they would fit together. We built the thrusters, the engine rooms in the belly of the ship, the circular hull with the crowning bridge on top. We constructed them to scale and planned how they would come together before actually assembling them into the starship. My dad was teaching me how to think like an engineer. A picture from a *Star Trek* poster served as our schematic, the Legos our components.

My dad started bringing home schematics from work. I had graduated to building replicas of the various engineering projects he was working on. Sometimes he brought home components from the industrial machinery and would explain the function of the component in the larger machine. He described how the component itself was made up of smaller components. I watched in awe as he turned the gears and explained their function. Then I built replicas out of Legos. These models actually functioned. We built interlocking gears and moving pistons, all of it hand powered, of course. But they would perform the same function as the real machinery, just at a much smaller scale.

I was beginning to understand how problems could be divided into smaller parts. I was seeing how *solutions* were made up of smaller, constituent steps. This kind of creative exploration was what we now call *hacking*. It was hands-on learning by puzzling through a problem.

These formative experiences instilled—or maybe just revealed—my passion for problem-solving. It was a passion that would stay with me throughout my life. I would later follow in my dad's footsteps and become an engineer myself. But I didn't know that at the time. I didn't know that I would someday apply the fundamental problem-solving skills he was teaching me to much bigger challenges in business, technology, and healthcare. At the time, I was just having some fun with my dad and the puzzle in my hands.

Science Nonfiction

My mom didn't really share my deep interest in puzzles, but she was supportive of it. She was a schoolteacher and believed in the value of learning, but building stuff wasn't her thing. We spent our time together socializing and watching television. I grew up in Hyderabad in southern India, but we got American shows on the local stations. My mom would sit and watch *Star Trek* reruns with me. She liked the adventure plots and the drama between characters. I was far more interested in the space exploration and futuristic technology.

The gadgets and machines employed by Kirk and the rest of the crew were fascinating. Speculative technologies like phasers, teleporters, and warp drives captivated my imagination. It was magic rendered in the language of science. What separates science fiction from fantasy is its plausibility. No one will ever be able to cast a spell to launch fire from their fingertips. But it is easy to imagine wearable bionics that would do just that. Being a pragmatic person, I have always been more drawn to the plausible.

I was so pragmatic that my daydreams about these fantastical speculative technologies extended into the mundane. I imagined *Star Trek* technology put to use in everyday life. I imagined being dematerialized and rematerialized on the other side of the Earth or flying through space at warp speed on a daytrip to the moon. Such technologies would radically transform humanity, if only they existed. As a kid growing up in the seventies and eighties, I was excited for the future. None of these technologies existed at the time, but it was thrilling to imagine that they someday might.

That future ended up not being as far off as I had imagined. While we are unlikely to ever see technologies that violate the laws of physics, rendering faster-than-light travel unlikely, many real-world modern technologies are analogous to what I grew up watching on *Star Trek*. By the time I was in college in the nineties, some were already hitting the consumer market.

The most obvious example is the parallel between cell phones and the *Star Trek* communicators used by the crew of the USS *Enterprise*. In 1996, Motorola released the original StarTAC phone, which became one of the first cell phones to see widespread consumer adoption. Motorola would go on to sell sixty million units before retiring the line. There had been earlier consumer cell phones that didn't catch on. They were large and clunky like early wireless house phones. The StarTAC was different. It was a flip phone that fit easily into a pocket or the palm of a hand.

The StarTAC flip-top design was remarkably similar to *Star Trek* communicators, which had a combination microphone/earpiece that flipped up. This was no mere accident, nor was the product's name. Martin Cooper led the team that developed the StarTAC. He was known for having pioneered early mobile phone technology in the 1970s. It had always been Cooper's dream to build a handheld mobile phone that functioned like the communicators in *Star Trek*. He actually cited the show as inspiration. "That was not fantasy to us," Cooper said about the *Star Trek* communicators when talking about his work on early cell phones, "that was an objective." The StarTAC was the *Star Trek* communicator's namesake.

Cooper wasn't the only person inspired by the speculative technology of *Star Trek*. The Altair 8800, the first commercially successful personal computer, is rumored to have been named after a reference to a *Star Trek* episode in which the crew visits the Altair star system. In fact, many technologies were inspired by other works of science fiction as well. Scientists and inventors have always taken inspiration from creatives exploring the bounds of the imaginable. Jules Verne's

1870 novel *Twenty Thousand Leagues Under the Sea* predicted many of the features of modern submarines in a time when their extant predecessors were primitive vessels. It was real-life engineers who made those visions a reality.

Cell phones aren't the only modern technology resembling the then-futuristic tech on *Star Trek*. Much of the speculative technology on the show is already a reality and commonplace. Video conferencing is no longer limited to the USS *Enterprise*'s bridge. Now you can get a Zoom chat room or video call a friend on Skype. Tablets like those featured on the original *Star Trek* series are now common consumer items. Most people in the developed world carry around cell phones that act like the PDAs common to science fiction. Tricorders, the scanning devices carried by the USS *Enterprise* crew, resemble handheld scanning equipment now in use in a variety of industries, from healthcare to mining to construction. Wearable computing devices, sensors, and screens are common in both the consumer market and in industries.

There are also many developing technologies that are starting to resemble speculative technologies. The *Star Trek* series famously papered over the plot hole that the language barrier between alien species would present by introducing universal translators that automatically translated a speaker's words for the listener. While no present technology operates so seamlessly, advanced voice recognition and text-to-speech software do allow people to communicate with each other in different languages on the fly. It's not hard to imagine how such technology could someday be miniaturized and implanted into the human body. There is no Holodeck, which appeared in later series of the franchise, but consumer virtual reality headsets are getting better and more immersive with each generation. The VISOR worn by Geordi in *Star Trek: The Next Generation* allows a blind man to see, not so different from cochlear implants that already allow the deaf to hear. Even now, scientists are racing to develop the bionic eye.

Some speculative technology hasn't materialized, of course. Technologies that break the laws of physics are unlikely to ever be real. We probably won't ever see warp drives that allow us to travel faster than light, transporters that move matter across space magically, or replicators that create prepared food from thin air.

But many technologies that were once thought impossible have become available as everyday consumer goods. There is reason to believe that at least some of the things we still think impossible may one day become reality. There are no tractor beams, not yet, but we can imagine all kinds of futuristic solar sails that move objects with lasers. General artificial intelligence is currently nowhere near the level of sophistication as seen in the android character Data, but given the exponential rate of advancement in AI, we can imagine a day when software might become sentient.

Exponential Technology, Exponential Change

More breakthroughs *are* on the horizon—and sooner than we might think. The pace of many technologies is exponential. They aren't growing at an incremental rate. Progress is compounding and growing faster and faster over time. We call these *exponential technologies* because they advance ever more quickly as they mature. Their growth follows the upside of a parabolical curve.

The exponential technology most familiar to the general public is computer microchips. Their exponential growth has been popularized by Moore's law. In a 1965 paper, Gordon Moore, cofounder of Fairchild Semiconductor and Intel Corporation, observed that the number of transistors on a microchip had been doubling annually for a decade and would continue to do so for another ten years. In 1975, again looking forward, Moore predicted that the same doubling would occur about every two years for the foreseeable future. Moore was making a bold prediction.

Despite the name, Moore's law is an observation and a projection, not a hard rule. Nonetheless, Moore's law has held true for over half a century now. Computer chips have doubled in power about every two years, give or take, ever since. They continue to do so today.

Perhaps almost as amazing is that as microchips have been doubling in power, they have similarly fallen in price at the same rate. Microchips have been getting exponentially cheaper as they have grown exponentially more powerful. This has allowed for a parabolic expansion in computing power. The processors in modern cell phones are now orders of magnitude more powerful and cheaper than the supercomputers of Moore's time.

Computing power is not the only exponential technology. Another key exponential technology is so-called big data. We have seen a parabolic explosion in the amount and quality of data over the last few decades. The world has deployed a massive physical infrastructure that collects data for various processes and operations throughout the global economy. They also pick up personal data. Today we have sensors in cars, smartphones, watches, monitors, medical devices, industrial machinery, appliances, light bulbs, radios, supermarket and postal scanners, red lights, doors, underneath highways, and in everything else as well. The Internet of Things (IoT) comprises billions of connected devices, with more being added ever more rapidly. In the next few years, we will surpass a *trillion* sensors operating on many billions of devices around the world. These sensors are providing data on anything and everything.

This expansion of big data will not be slowing down anytime soon. The world is producing more data than ever. It is hard to overstate how much data now exists. We have created more than 33 zettabytes of data as of 2018. Market intelligence company IDC estimates that the world's cumulative extant data will grow to 175 zettabytes by 2025.[5] This is a number that the human brain cannot even comprehend. The average 25 mb/second broadband internet connection downloading this much data full-time would take nearly two billion years to finish.[6] Much of this existing data was created in the last few years. That's the nature of exponential growth. You're *always* riding the steepest part of the curve—but it just grows bigger and bigger.

This data is not just Netflix streams and TikTok videos. Much of it is consumer and industry data. In the healthcare industry, the total amount of information doubles every seventy-three days, according to a 2011 study put out by the *Transactions of the American Clinical and Climatological Association*.[7] This is due to the increase in medical records, images, labs, and other medical data, all of it now being recorded digitally. This vast amount of information provides more powerful insights than ever before.

The exponential growth in data and processing power is driving exponential advancement of several other technologies too. Technologies don't exist in vacuums; they reinforce one another. Exponentially

5   Oliver Peckham, "Global DataSphere to Hit 59 Zettabytes in 2020 Alone, IDC Projects," Datanami, May 19, 2020, https://www.datanami.com/2020/05/19/global-datasphere-to-hit-59-zettabytes-in-2020-alone-idc-projects.

6   David Reinsel, John Gantz, and John Rydning, "The Digitization of the World from Edge to Core," International Data Corporation, November 2018, https://www.seagate.com/files/www-content/our-story/trends/files/idc-seagate-dataage-whitepaper.pdf.

7   Peter Densen, "Challenges and Opportunities Facing Medical Education," *Transactions of the American Clinical and Climatological Association* 122 (2011): 48–58, https://www.ncbi.nlm.nih.gov/pmc/articles/PMC3116346.

increasing computing power and data sets allows for the modern artificial intelligence systems that we have today. In fact, these advanced AIs are required just to crunch and process all the data.

These AI systems are able to improve all kinds of other technologies as well, including artificial intelligence itself. AIs can process their own data to improve their own performance, which is the basis of machine learning. This is just one example of how exponential technologies feed off each other and snowball ever faster.

Not all technologies are accelerating exponentially, but many are. Anything with a microchip is potentially an exponential technology. Not just computers but also smartphones and other devices. The number of sensors on a smartphone has been doubling every four years. The IoT can itself be seen as an exponential technology, expanding to encompass all physical infrastructure. Currently, the growth in genetics and bioengineering is undergoing exponential growth with the advent of technologies like CRISPR gene editing, which rely on processing vast amounts of genetic data. Cloud computing and streaming services are growing rapidly in usage and sophistication. Most technologies that benefit from more processing power and data are developing at a faster rate right along with these drivers of exponential growth.

In many cases, the adoption itself makes the technology more useful. More people and businesses adopt technologies as they improve and mature. At a certain point, adoption becomes its own driver of exponential growth. More companies and people using the technology actually makes it inherently more useful. Consider the network effect of the telephone or, more recently, the internet. The more people who own phones or have internet connections, the more useful they become to the user. This in turn drives further investment, which makes the technology even more useful, which leads to ever more adoption. This self-reinforcing cycle eventually sets off an exponential explosion in the utility of a technology. The more adoption, the greater the utility. This is especially true for exponential technologies. Consider big data. Facebook and Google are so valuable because they produce data from so many users. More data means more and better insights.

This effect is not limited to social media data. In the healthcare industry, widespread adoption of new technologies also produces more useful data sets. More equipment means more sensors, which mean more data. This allows medical professionals to glean more and better insights. Doctors can better understand patient populations. Insurers can develop better insights on claims and processes. Hospitals can better understand all of their metrics. And all of this value compounds and increases exponentially.

Although it begins with a trickle that grows over time, the exponential growth of technology can appear like a sudden explosion when it first bursts onto the scene. This is usually an illusion. Exponential technologies grow slowly at first, doubling and doubling their speed over time until they mature to a certain level of commercial viability. This is when mass adoption can make it seem like they came out of nowhere. In truth, it was exploding in growth all along, and no one noticed until it swept the world.

This is the nature of exponential expansion, but it can be hard for even intelligent and highly educated people to grasp on a gut level. Consider this thought experiment: An invasive species of water lilies starts growing in the middle of a large pond. They reproduce rapidly, doubling every day. If it takes them thirty days to fill the pond, how long does it take them to cover half the pond?

The answer, of course, is twenty-nine days. This is easy enough to reason out, but the answer *seems* wrong. That is because people think of growth in linear, incremental terms. We have a hard time grasping exponential growth. It is easier to imagine the effect of saving $500 a month for forty years than the much larger effect of compounding interest over time. We might think the monthly deposit matters most, but in truth, the average portfolio compounding at an average of 8 percent interest will grow the portfolio to six or seven times what it would otherwise be. The effect of exponential growth absolutely dwarfs the incremental savings.

Where the pond and investing metaphors break down is that, in the case of technological progress, there is no pond to limit growth, nor is there a life span limitation. Exponential technological growth can theoretically continue on forever.

This is why technological progress can feel like it is always exploding just now, like we are always on the edge of something bigger—because we are! The advent of computing was a huge breakthrough. So was every step of growth along the way. Personal computers were just as revolutionary as early supercomputers. Smartphones are revolutionary. Quantum computers will be revolutionary. Each doubling has a larger and larger impact. This is hard to grasp looking forward but easy to see when looking back. We just can't imagine what technology has in store for the future.

I have zero doubt that the exponential pace of technological development ensures that those living today will see technology and scientific breakthroughs once unimaginable outside the bounds of science fiction. The speculative is becoming reality—and doing so at an ever-accelerating rate. This has been the case for much of recorded human history. What is exciting about recent times is that the pace of progress has accelerated to the point where major technological and societal shifts unfold over the span of a single generation, over a single lifetime.

People born on the cusp of the industrial revolution saw massive change in their lifetimes. They witnessed the rise of automation, steam rail, and countless other industrial marvels. Those born in the early twentieth century witnessed even more mindboggling technological advances. They may have grown up with horse-and-buggy carriages, seen the country urbanize and then suburbanize as car culture proliferated, and spent their golden years zipping across the country on commercial airlines. People born before the invention of the first electronic television in 1927 are using smartphones today.

The exponential acceleration of technological advancement continues now. Those of us living today will be shocked by the scientific breakthroughs in the coming decades. Gene editing, nanotech, biotech, advanced artificial intelligence, quantum computing, and other exponential technologies are set to change the way humanity lives and conducts business. The coming change will not be just as revolutionary but actually *more* revolutionary than the change previous generations witnessed.

If science fiction has shown what humans can imagine, science has shown what we can accomplish. However, it is business and industry that ultimately bring novel technologies to market, put them to use, and refine their application over time.

CHAPTER 2

THE BUSINESS OF EXPONENTIAL TECHNOLOGY

HEARING ABOUT HOW *Star Trek* communicators inspired Martin Cooper to develop the cellular phone was an aha moment for me. His inspiration was my inspiration. I, too, was inspired by the technological possibilities put forward in *Star Trek* and other science fiction classics. Moreover, I was inspired by Cooper himself. Here was a man who took speculative technology and *actually built the thing.* He didn't just marvel at fantastical speculative technology—he asked, how would we build that? Then he figured out a way to do it for real.

And the thing *worked.* He built a working communicator! Thanks to his efforts, and those of countless others, we can now talk to people on the other side of the world in real time, on the go, from a device that fits in our pocket that the average person can afford. Science fiction became science reality.

From Inspiration to Methodology

This wasn't easy, of course. Inspiration alone isn't enough. You still have to figure out how to build the idea. Science fiction can provide inspiration, but scientists and engineers have to do the hard work of discerning what is possible with present technology. You have to go from inspiration to methodology and practice.

Cooper showed me that there was a realistic method for turning fantastical ideas into real-world inventions. That was his life's work—and I wanted it to be my life's work too. I wanted to make the impossible possible. I already had the inspiration. The possibility of novel technologies had always been inspiring to me from my first episode of *Star Trek* as a child. I have stayed as inspired by the great possibilities of science and technology ever since. The methodology—I was still working on. I had a natural inclination for problem-solving, but the problems I wanted to tackle weren't simple puzzles. They would take training and experience to practice effectively.

My journey, as it does for most technologists, began with a technical education. I majored in mechanical engineering at a college in India before moving to the United States to pursue a graduate degree in computer science at the University of Pennsylvania. This was around when I had my aha moment about Cooper, which was confirmation that I was on the right path in life.

The switch from engineering to computer science was as seamless as the jump from jigsaw puzzles to Lego blocks. Software engineering is not so different from other kinds of engineering; it's all problem-solving. Software is built the same way as any other engineering solution. A problem is identified, solutions are planned, and then you have to figure out the best way to build the idea. Good architectural principles apply to code just as they do to other engineering problems, such as, well, architecture. You build a program the same way you construct a building. Good code is made of discrete interlocking parts that snap together to form a working whole. It's no different in concept from putting together Legos or puzzle pieces or mechanical components. For me, software was just another puzzle to solve, but in some ways it was purer. The virtual world of computer programming makes real-world problems abstract, distilling them down to their purest concepts. But the problems are often similar. In a way, most problems are.

So why choose code? Because it was an exciting time to get into programming and software. I could see that this was a growth technology. I didn't necessarily think about it in those terms, but I could feel it. Personal computing had matured into viability. Most middle-class families now had their own personal computers in the home. It was also the early days of the internet, which had become commercially viable. Dial-up internet was everywhere. The whirr of dial-up modems and "You've got mail!" notifications were sounding out across the country.

The dotcom boom was just over the horizon, but already there was an excitement about the internet. Everyone in business, certainly in Silicon Valley, was waking up to the potential of online commerce. Traditional big businesses were taking notice, as were newcomers. Google, Yahoo!, and Amazon were still just start-ups at the time. In the Valley, it seemed like everyone and their cousin were working on a dotcom venture. Commerce was moving online—we all knew it—but we still had to fumble our way to the right business model. This is the way it is with all new tech. Online shopping seems commonplace now, but at the time, it was truly revolutionary. We were figuring things out as an industry.

I wanted in on that action. I saw a chance to use this new technology to solve everyday problems. After working as a developer for a few tech companies, I launched my own e-commerce start-up. Most online retailers at the time had a niche. Amazon started as a humble online bookstore. Our niche was online furniture shopping. We wanted to leverage the internet to put custom-made furniture at the consumer's fingertips.

The website we were developing allowed customers to choose, build, and customize furniture during the production process. They could watch and make changes as the pieces moved through the assembly line. The website would provide real and virtual images of the piece, allowing them to swap colors, features, and sizes on the fly and get a customized piece of furniture delivered within a few weeks. This may seem standard or routine now, but it was novel at the time. None of this was possible without the widespread adoption of home computers and the internet.

The process was partially inspired by all of those Legos sessions with my dad. Customers could plan and assemble their furniture online before we built and shipped it to them. Or that was how it was *supposed* to work anyway. Unfortunately, the company fell apart before the website ever went live. The dotcom boom gave way to the dotcom bust. Venture capital and other funding dried up. We didn't have the money to finish developing the website, much less any production capacity. We were forced to shutter the project, sell off assets, and move on to other things.

Despite our inspiration, the company hadn't worked. We had gotten something wrong—many somethings, really—and failed to come up with the right business model at the right time. But in doing so I had learned a lot about the business of business. The lessons learned would not be for naught.

Learning the Business of Healthcare Technology

For my part, I went back to working as a technology consultant for several technology companies and consultancies over the years. This was how I got into the healthcare space—by way of happenstance. One of my first big clients was a large integrated health system. I spent five years helping them modernize their various systems to leverage new technologies.

This experience gave me a crash course in the entire healthcare industry. The health system was a consortium of primary care providers, hospitals, and health insurers united as one mostly self-contained ecosystem. Their business model was unique in this way. They had their hand in practically every part and corner of the American healthcare system; it was the American healthcare system in microcosm. I got to see how all of the puzzle pieces fit together. The healthcare system is really several systems that interlock in various ways. Each of those systems has subsystems, and those subsystems have individual parts with their own components. It's an incredibly complex machine that makes up nearly a fifth of the American economy.

Being able to see and work with the whole American healthcare system in miniature was invaluable. I could recognize not only how the different components worked together but also the many points of friction. As a consultant helping this health system modernize its systems, I learned about the challenges facing the healthcare system as a whole. I learned what worked and what didn't and what points were under pressure and strain.

For all its strengths, the American health-care system has real problems.

> **The tech is just the tool. Knowing how to use the tool isn't enough. You also have to understand the business context in which it is being applied.**

When I later went to work in Deloitte Consulting's healthcare technology division, I saw that my observations about the health system did indeed apply to the wider healthcare industry. I worked with all kinds of providers, insurers, hospital and care centers, and all of the other actors that make up the American healthcare system. This confirmed what I had already learned: for all its strengths, the American healthcare system has real problems.

This was concerning, of course. But I wasn't just an observer. I was a consultant working to solve problems. I was a *technology* consultant, in particular, which meant that we were using new tools to solve long-standing problems. Yes, there were problems. But we were finding solutions. We were leveraging technology to do things in new and better ways.

Solving such problems isn't just a matter of having the tech, though—that's just the precursor. These weren't tech problems. They were business problems to be solved with the application of technology. Many tech people fail to internalize this distinction. The tech is just the tool. Knowing how to use the tool isn't enough. You also have to understand the business context in which it is being applied.

My years as a technology consultant and now as a healthcare executive have made this truth abundantly clear. It was easy to see as a software developer. Knowing a programming language allows you to build solutions that can scale. But building the *right* solution in the most efficient and effective manner requires a deep understanding of the business context. Professional software development taught me how to examine problems through a business lens. What do the stakeholders, customers, and staff need? What problem are you trying to solve? How will a new implementation improve a business process?

Tech for the sake of tech is a wasted investment and a missed opportunity rolled into one. The newest tech won't help if it's not the right tool for the job. A fancy pneumatic staple gun may outperform a manual stapler, but it won't help you drive a screw into wood as well as a cheap screwdriver. You can't apply eyeshadow with a tube of lipstick. And you can't build optimal digital solutions with the wrong tech.

The more exciting the technology, the easier this truth is to forget. This was where many companies went wrong during the dotcom era. Those were heady times in Silicon Valley. There was a utopian ethos. We felt that technology was about to radically transform society. This sentiment so permeated the industry that you could feel it in the air. You saw it on other people's faces. You saw it on your own in the mirror. There was a boundless sense of optimism that technology— and we as its purveyors—could transform the world. Any business with a "dotcom" at the end of its name seemed destined for greatness.

Of course, this turned out to be folly. Many early tech companies turned out to be smoke and mirrors—and they often didn't even know it at the time. They just didn't really understand the business context in which they were operating. They lacked a realistic business plan. They didn't focus on how the internet could improve business processes or create better ones. Many companies failed, my own among them. Pets.com became a punchline and the posterchild of overhyped tech companies without a viable business.

The skeptics got the last laugh. They had been asking all along: What's the business model? How will you make money?

When there had been a seemingly endless stream of capital flowing into dotcoms, these questions had seemed irrelevant. It was only when the easy money dried up that companies were forced to be pragmatic. Not every growing company necessarily had to turn a profit right away—but they had to have a realistic plan for being profitable someday and another plan for getting there. Many companies simply did not. Those were the ones that failed.

Other companies survived and thrived. They built viable businesses that transformed the economy and our lives. These successful companies weren't just using the latest tech; they were deploying the right tech under the right business model. The companies left standing when the dust cleared were those founded on fundamental business principles. They were solving business problems and answering consumer needs.

This doesn't apply just to tech companies—it applies to all companies. That's because all major companies use and leverage tech, no matter their business. This is certainly true of the healthcare industry. In the modern world, every business process is touched by technology. It took me five long years to help modernize Kaiser Permanente's dated technology systems. (This modernization process is ongoing, as it always is, but we made leaps and bounds in that time.) Some records were still only on paper. Their electronic databases weren't streamlined or well integrated. Many business processes that could have been automated were not. The entire consortium needed to change how it conducted business at every level.

This was a massive undertaking that involved far more than just setting up some stock enterprise software. They needed custom solutions. That requires a deep understanding of the business context, not just the technologies. Technologies interact with business processes. Bringing in new tech and adapting it to the current way of doing things is suboptimal, even counterproductive. We have to go through each and every business process and consider if and how it could be improved with new or better tech. We also have to consider how each business process might need to change. New tech calls for new ways of doing things. Tech must be seamlessly integrated into business processes. The two are inextricable because it's a business problem you want to solve with tech. The right solution blends the tech into the business process.

Then there's the matter of adoption and compliance. The entire consortium was being asked to change how it conducted business. Kaiser Permanente was transforming how it handled information, interacted with customers, underwrote and sold the businesses—everything about the organization was up for overhaul. While the result would be a better and more efficient system, getting there required a huge shift in behavior and mindset. Staff, providers, hospitals, insurers, and everyone else would need to adjust to new ways of doing things. Even patients, customers, and vendors would need to adapt to the change—and it is on the company to help them do so.

The human element of adopting new technology is critical. Sudden changes can cause discord without proper change management. In order for the new systems to work, everyone has to carry them out properly. Otherwise, expensive new tech spending goes to waste. Networks and integrated databases won't achieve their full potential unless they are fully integrated with other systems and utilized properly. The time providers spend inputting data into a database goes to waste if no one is actually using the system for anything else.

It isn't just staff who has to be conditioned to big changes—managers and executives resist change too. The first job of a technology consultant is usually to convince leadership of the necessity of radical changes. Leadership is often most afraid of rocking the boat when it comes to long-standing business processes that have worked (well enough) for them so far. For all the talk of flexibility and nimbleness in business, executives and management are very resistant to change, especially big change. With so much money on the line, leadership is often biased toward the familiar, the tried and true.

Case in point: At a client of the previously mentioned health system, the head of customer service pushed back on our recommendation to implement an automated online customer service system. This would save money by reducing the traffic to customer service phone lines that had to be staffed by live people. The head of customer service was skeptical. He didn't think anyone would trust an automated service. He worried it would actually increase the burden on the customer service lines when people called in to verify what they had read and done online. The customer service reps would then be forced to explain the system on top of everything else.

This may seem preposterous now, but at the time, these systems were new. No one had much experience with them. We had to convince them that the system was a better way of doing things. This isn't always easy. In this case, we had to show the head of customer service a working prototype. We sat him down at a computer and gave him a phone. First, we had him call in for information. Then we showed him how easy it was to navigate the system. This convinced him. There was no calling in and getting passed around, no explaining things over the phone, no reading off numbers to the customer service representative.

"I get it now," he said.

He was convinced. But it took a hands-on demonstration of how the technology improved a major business process to do so. This slowed down the modernization process, but his concern was legitimate. New tech is only worth implementing if it actually improves a process and adds value. Shrewd business leaders only care about how tech will make their processes more efficient and their businesses more profitable. This is the correct mindset, so when implementing change, there should be a strong case for exactly how this new technology will actually improve business or operations.

With many exponential technologies, as long as you are applying them to the business context properly, this is not hard to do. The improvements we made to the customer service system were substantial. And as the technology got better, the system improved further. Today, many healthcare companies employ AIs and chatbots to assist customers. Due to the exponential advancement of these technologies, they are getting better all the time. No one questions them anymore.

That's why exponential technologies are so valuable to businesses—properly applied, exponential technologies get better and better, faster and faster.

Timing the Parabolic Curve

Over my decades-long tenure in technology consulting, I have seen many technologies come and even go. I won't be out of work anytime soon. There is always some new technology coming down the pipeline. What stays the same, year after year, is the process of integrating the newest technologies into business processes and structure. This is why businesses need technology consultants. They are not hired for being on the cutting edge of technology. With truly novel tech, they are often learning the tech right alongside their clients. Their true value is in knowing how to help companies restructure their businesses around new technology.

These "new" technologies often aren't actually that new or distinct. They tend to be new iterations of or forks within existing exponential technologies. The internet boom in the nineties was really an expansion on existing networking technologies. Bulletin board systems (BBSs) offered a way for millions of users to dial up shared servers to access online forums long before cheap dial-up internet service. The internet only seemed novel because it was so much better than BBSs and therefore enjoyed greater mainstream adoption.

This caused an exponential increase in interest in the technology, including among the private sector and business communities. Once the internet achieved a sufficient level of advancement and adoption that e-commerce and online advertising became viable, the dotcom boom started in earnest. The exponential curve steepened further, and investment took off.

We see the same pattern in all exponential technologies. And we are seeing it now with artificial intelligence and blockchain, the primary subjects of this book due to their not fully tapped ability to revolutionize American healthcare. Neither of these are new technologies. The first proper blockchain was the Bitcoin cryptocurrency blockchain introduced in 2008. But work on cryptographically secured chains of data "blocks" was being discussed in the early 1990s. The technology has just grown exponentially over time ever since. Artificial intelligence had more of a head start. It dates all the way back to the 1940s when Alan Turing used rudimentary AI technology to break the code being used by German ciphers during World War II. From there, artificial intelligence grew exponentially for years, getting more and more sophisticated.

It is when exponential technologies cross some threshold of usefulness in business processes that they only appear to suddenly come out of nowhere. This is an illusion, in a technical sense. They have been growing exponentially for years. It is just that some new iteration or application reached maturity and commercial viability that makes the exponential curve seem to lurch.

We have seen this cycle play out with artificial intelligence over the last decade. The catalyst was a 2011 appearance of IBM's Watson on a special run of *Jeopardy!*, the popular American gameshow. Watson was an AI computer system IBM created to answer questions posed in natural language. The system was designed specifically to compete on *Jeopardy!*, but its ability to interpret natural spoken language and synthesize answers from massive databases had obvious implications for all kinds of public and private business uses.

Watson faced off against Ken Jennings and Brad Rutter—the first and second highest-earning *Jeopardy!* contestants of all time, respectively—and came away with the $1 million first-place prize. Watson's appearance on the gameshow reignited business interest in artificial intelligence. It had made it clear that AI was coming into a new stage of maturity. The new business potential was clear.

IBM marketed Watson heavily to all kinds of organizations with a particularly heavy focus on the healthcare sector. The promise of Watson convinced many companies and organizations to go all in on artificial intelligence. Many shot straight for the moon.

Unfortunately, in doing so, many missed.

Perhaps the most infamous and spectacular of these failures was a 2013 joint project between IBM and the MD Anderson Cancer Center at the University of Texas. The center already had an ambitious mission: the complete eradication of cancer. They applied IBM's technology just as ambitiously. They launched a pilot program, Watson Health, that would feed the Watson cognitive learning system all available cancer research and patient data. They were teaching it to diagnose and treat cancer patients. The hope was that it could synthesize patient data more thoroughly than human doctors. Four years and $62 million later, MD Anderson pulled the plug on the project with little success. Other organizations suffered similar failures with Watson as well.

So, what went wrong? Why was Watson able to best some of the world's greatest trivia champions but couldn't solve basic business problems?

As it turns out, teaching an AI to play *Jeopardy!* is a fairly simple problem to solve. The researchers just needed to feed Watson a massive amount of data—far more than Watson Health was ever given—and have the system return an answer when it found one with a reasonable probability of being correct. This worked well for winning trivia games. Watson was able to process and synthesize massive amounts of data faster and more accurately than humans. It was also able to issue answers within a confidence certainty that allowed it to win more than it lost.

However, in the case of cancer research, there is actually limited data. Patient records could only be matched to so many relevant studies. The problem for Watson Health wasn't too much data—it was too little. Just as problematic for the project is the life-and-death nature of making medical diagnoses. The stakes are low in a trivia game. Watson was able to ring the buzzer quickly and spit back the right answer with a reasonably high success rate. It got some wrong, but it answered enough correctly to win the game. This doesn't work in medicine, where a misdiagnosis can mean death. Although Watson sailed to victory easily in *Jeopardy!*, MD Anderson wasn't able to trust Watson Health enough to make diagnoses on its own.

Watson also proved unable to make major business decisions. Businesses tried, but Watson's insights were too faulty to be trusted. In business, you might not always have lives on the line, but you do have livelihoods on the line. When millions or billions of dollars are at stake, you still cannot afford for the computer to make errors. This limited Watson's application in business. The system wasn't able to make superintelligent decisions, and thus businesses were not able to employ Watson in the ambitious ways that they had hoped. The problems were just too complex for the AI to solve reliably.

The lesson here is not that AI cannot be applied to business problems. The lesson is that you have to be careful *which* problems it is applied to and how. The technology has to be capable of solving the problems it is given. You have to consider the business context and the particular set of constraints you are operating under.

Although Watson Health was derided in the media as an oversold failure, the reality is that many organizations and companies were able to employ artificial intelligence to solve business problems. AIs were even able to assist in making better business decisions. I know because I was involved in deploying the technology in this way for many healthcare companies and institutions.

I was still working with Deloitte at the time. One of our main clients was Anthem Inc., an innovation leader dedicated to improving health and probably best known for its affiliated health plans. We watched Watson clean up on *Jeopardy!* and wondered—like so many others—what kind of business problems such an advanced AI could help solve. There was only one way to find out. We applied artificial intelligence to various problems. Many proved unsuccessful, as Watson Health had been, but others worked.

The most successful projects were those limited in scope. We were able to use AI to automate several processes that were previously performed by staff. Natural language processing allowed us to gather data from unstructured documents and draw insights without anyone having to take the time to read them. AIs proved effective at interpreting clinical documentation. We automated more of the underwriting process. This isn't that far from what MD Anderson had attempted. We succeeded where they failed by limiting the role of the system, which would make decisions on claims within a certain confidence interval and flag edge cases for human review. This lowered the risk of false positives or negatives while still automating most of the claims process.

These steps worked because they were targeted, measured, and practical. We tempered our expectations and acted pragmatically. In doing so, we were able to help Anthem improve many processes in measured but real ways. Ambition is good. Being overly ambitious is not.

> The trick to making the most out of exponential technologies is timing. New technologies should be applied to the problems they can solve now. Perfectionism leads to missed opportunities.

A few years later, I had the privilege of joining Anthem. My role was to help Anthem leverage AI and other exponential technologies to transform it into what we call an *AI-first company*, by which we mean a company that embeds AI into the decision-making process at every level of the organization. Over the next few years, we continued to implement small projects with great success. We were always careful about how we chose to implement artificial intelligence. We didn't want to overshoot what was possible.

This measured approach was not borne of reservation with or doubt about the technology and its potential. We very much believed in the artificial intelligence. We could see the technology working. We were just practicing a pragmatism rooted in our understanding of business fundamentals and the current limitations of the tech. We maintained a business-first approach that was results oriented. Again, we made sure that we were applying technology to actual business problems. Anthem had been convinced to go all in on artificial intelligence precisely because of the efficacy of this measured approach. We can quantify its effect on business. We were seeing steady success operating in this way.

The trick to making the most out of exponential technologies is timing. New technologies should be applied to the problems they can solve now. Perfectionism leads to missed opportunities. I spent so much time perfecting my e-commerce business that we never got it to market. But overzealousness leads to disappointment, disillusionment, and—ultimately—disinvestment. It was true of the dotcom era. It is true of artificial intelligence now. It will be true of the next exponential technology as well.

Shooting for the moon only to miss will disappoint investors and early movers when it wastes resources and yields poor results. Applying exponential technologies to their maximum ability, but not beyond, allows you to string together a series of meaningful successes. This raises acceptance and adoption and maximizes the *current* potential of the technology. From there, you can tackle more and bigger problems as the technology matures and you ride it up the exponential curve.

Of course, this raises the question: How do you know when you are surpassing a technology's current ability? The answer, which we will explore later in this book, is through ample experimentation with a rapid and reliable experimentation infrastructure. Watson was a failure only in that expectations were too high, but the actual process of building that early prototype helped identify the current constraints of the technology. As a prototype, Watson was a failure. As an experiment, it was a success.

Someday artificial intelligence systems will be able to know everything and do anything. They will certainly be able to give better cancer diagnoses. There may be no limit to what they can predict. There may be no limit to what they can puzzle out. That's where the technology is headed. But that's not where we are *now*. In order to get there, we have to keep climbing the curve. That means applying exponential technologies now in pragmatic ways that fulfill current use cases. We can do a lot with what we have now.

In the healthcare industry, we can implement current technologies to improve business models, build more robust health systems, and achieve better health outcomes now. The healthcare sector is facing a number of problems that we now have the technology to solve.

Let's take a look at them.

CHAPTER 3

THE HEALTHCARE PUZZLE: A SYSTEM UNDER STRAIN

MAKE NO MISTAKE: The United States has the most advanced large-scale healthcare system in the world. Our hospitals and other medical facilities are world class. We are home to the top research institutions and pharmaceutical companies. Together, they are conducting the best research and developing the most cutting-edge therapies anywhere. In terms of research and development output and high-tech care, the American healthcare system is second to none.

Even the oft-criticized American health insurance system, though not without its faults, is among the best in the world by many measures. Our public/private hybrid combines free-market efficiency with many of the resiliencies and safety nets of a public healthcare system. Medicaid and the subsidies provided by the Affordable Care Act help insure approximately eighty million Americans who might not otherwise be able to afford healthcare. Another sixty-five million American seniors are on Medicare, which is also a public/private hybrid. However, most Americans get excellent private health insurance through their employer. Employer-based health plans are often subsidized and comprehensive. This rich patchwork provides health insurance to roughly 90 percent of Americans, underwriting the cost of the most advanced healthcare system on Earth.

Despite these strengths, the American healthcare system has much room for improvement. We are not without problems in need of solutions. Despite being the most expensive healthcare system in the world, the United States routinely ranks far lower than other developed countries in terms of investing in healthcare: as low as twenty-seventh in the world in 2016, a dismal ranking for the most costly health system in existence.[8]

These rankings are, of course, subjective and dependent on the metrics evaluated. Millions of Americans lack health coverage, which drags down the ranking and care. This isn't the only problem, though. Many Americans who do have coverage still find healthcare costs increasingly unaffordable as premiums and co-pays rise to cover ballooning prices.

In this chapter, we'll look at how misaligned incentives across the healthcare sector create systemic inefficiencies and disincentivize investment in economical preventative care, infrastructure modernization, and other investments that would bring down costs to more manageable levels.

This is a multifaceted problem. The healthcare puzzle is not a single dilemma—it is a set of interlocking challenges that must be addressed in concert.

---

8   Stephen S. Lim, Rachel L. Updike, Alexander S. Kaldjian, et al., "Measuring Human Capital: A Systematic Analysis of 195 Countries and Territories, 1990–2016," *The Lancet* 392, no. 10154 (2018): 1217–34, https://www.thelancet.com/pdfs/journals/lancet/PIIS0140-6736(18)31941-X.pdf.

Misaligned Incentives

The fundamental problem at the heart of the healthcare system, from which nearly all other problems arise, is a series of misaligned incentives. In the last chapter, we saw that the healthcare industry is made up of many sectors and actors, most notably patients and policyholders, providers and medical institutions, and insurers.

These actors are not operating under the same motivations. The healthcare system incentivizes behaviors that lead to friction and inefficiencies throughout the system. Different actors are tugging in different directions. Policyholders engage in unhealthy behaviors that make them more expensive to insure. Providers benefit from offering expensive treatments instead of low-cost preventative care. Insurers are underinvesting in preventative care that would keep patients healthy while reducing their own bottom line. These kinds of inefficiencies are driving costs up and pushing the quality of care down.

> Misaligned incentives in the healthcare industry are unintended consequences of how the insurance industry underwrites healthcare costs.

This is not some moral failure on the part of the actors, not mostly. The blame doesn't lie at the feet of any one group. It's a structural problem. Misaligned incentives in the healthcare industry are unintended consequences of how the insurance industry underwrites healthcare costs.

In most other sectors of the economy, buyers pay sellers directly for goods and services. Payment is generally made at the time of transaction. For very large purchases, buyers may need a bank loan, but even still the price and terms are generally known. The buyer makes payment more or less directly to the seller (or directly to the lender for the loan). There may be long supply chains involved, but every link in the chain includes a buyer and a seller transacting directly.

Direct transactions keep the incentives of buyers and sellers of most products and services well aligned. Consumers have specific wants and needs. Companies are rewarded for meeting market demand at the most competitive price. This is the basis of market efficiency. In most industries, it works very well. Consumer demand is clear, and companies succeed when they meet that demand. The buyer and seller almost always find their interests aligned.

This simple market structure breaks down in the healthcare sector where *insurers stand between policyholders and providers*. Most consumer healthcare spending gets filtered through the insurance industry. Policyholders are the buyers—but they aren't paying sellers directly. The insurance company pays providers for most services rendered. The buyer and the payer are separate entities with separate interests.

Decoupling the buyer from the seller—as insurers necessarily do—creates a situation in which there is no reason for buyers to ration purchases. Policyholders are paying the same premium no matter how much healthcare they use—why wouldn't they use as much as possible? They don't care what the insurance company pays. They just want their claims paid.

The seller is also disincentivized to ration care. Pushing procedures that are of marginal utility brings in more dollars without risk of chasing away a buyer who is insulated from the true cost. The patient doesn't care what the insurance company is being charged.

Deductibles, co-pays, and coinsurance that introduce cost sharing at the point of purchase are meant to mitigate this disconnect between buyers and sellers, but the problem remains. Healthcare services are expensive. Coinsurance, deductibles, and even maximum out-of-pocket limits are often dwarfed by the actual cost of healthcare. Elderly cancer patients on death's door have no reason not to seek a $10 million experimental treatment with a 2 percent success rate if the insurance company is footing the bill. Providers have no reason not to perform needless or hopeless treatments when they get paid regardless of medical outcomes.

The most ham-fisted solution would be to do away with insurers entirely. This is not an option. Modern healthcare is expensive—too expensive for most people to cover out of pocket. Newly approved cancer drugs cost $10,000 a month on average.[9] The average hospital stay is also $10,000.[10] The average cost of having a baby is almost $11,000, even without prenatal services, aftercare, or delivery complications.[11] And these are just averages. The cost for particular individuals is often much, much higher. Emergency care or managing chronic illness would be prohibitively expensive for most people without insurance to rely on. Most people don't have a million dollars sitting around to pay for cancer treatment.

This is why we have insurers. They are not just middlemen. Insurers are in the business of underwriting the collective risk. The cost of healthcare does not disappear when it is invoiced to the insurance company. Someone still has to pay. Premiums are calculated based on the average cost of insuring a population of policyholders. When costs go up, so do premiums. Insurance companies couldn't remain

9   Jessica Merrill, "Remember When Provenge's Price Was Bold? Every New Cancer Drug in 2017 Cost $100,000 or More," PharmaIntelligence, accessed October 2020, https://pharmaintelligence.informa.com/resources/product-content/every-new-cancer-drug-in-2017-cost-100000-or-more.

10   Matthew Michaels, "The 35 Most Expensive Reasons You Might Have to Visit a Hospital in the US—and How Much It Costs if You Do," Business Insider, March 1, 2018, https://www.businessinsider.com/most-expensive-health-conditions-hospital-costs-2018-2.

11   Rickie Houston, "How Much Does It Cost to Have a Baby?" SmartAsset, October 1, 2020, https://smartasset.com/financial-advisor/cost-of-having-a-baby.

solvent without passing costs back to the policyholders. There's no such thing as a free lunch. More healthcare spending results in higher premiums for everyone.

The problem is that higher premiums don't incentivize judicious use of care the way that paying the actual bill would. If anything, higher premiums have the *opposite* effect, as policyholders paying exorbitant healthcare premiums might feel like they need to get the most bang for their buck. If you're already paying $2,000 a month or more for health insurance, you are going to feel downright entitled to all the care you can get. This creates a vicious cycle, driving premiums ever higher, as we have seen for many years now. Incentives aren't just misaligned—they're falling ever more out of alignment.

Not Enough Preventative Care

There are other misaligned incentives in the healthcare system. Immediate incentives are often misaligned with long-term interests. Disease prevention is always preferable to treatment or management. You cannot die from a disease you don't get. Preventative care is also orders of magnitude less expensive than treatment. Screenings, labs, checkups, and wellness spending are far cheaper than managing advanced or chronic illness.

Nonetheless, patients typically don't seek out care until they get sick. They are reactive to their health. They don't seek out care until they already have a health issue to address. Providers and insurance companies also place little emphasis on prevention. They sometimes even resist expansion of preventative care, even though it would bring down costs.

This is *not* a case of providers and insurance companies failing to act in their own interest. It's often the opposite. Providers make more money performing expensive medical treatments than they do from issuing preventative care guidelines. Every sickness avoided is an invoice not billed to insurance. Doctors aren't plotting to get their patients sick. They just aren't incentivized to take the active measures to keep them healthy. There's no financial downside in ignoring preventative care, at least not in the short term, so it's easier to just keep seeing patients as they get sick.

Insurance companies are also not greedy or evil. They aren't trying to drive up costs by avoiding preventative care. They are just reacting to market conditions. It often makes good business sense for any particular insurance company—when viewed in a vacuum—to not pay for preventative care. Market pressures always encourage short-term efficiency over long-term prudence. Preventative care is an easy place to cut corners without losing policyholders to competitors.

Even when an insurance company does invest in preventative care, there is no guarantee that they will see the return on their investment. Preventative care requires up-front investment to avoid much larger payouts down the road. Most people will switch insurance companies many times over the course of their lives. Insurance companies won't realize a return on preventative care if the policyholder switches to another insurer. The benefit goes to the new company. In paying for preventative care, insurance companies are subsidizing their competitors.

Insurance companies cannot easily address this problem on their own. The insurance industry faces a collective action problem here. Individual insurers find their own interests misaligned with those of the wider industry. If the whole industry invested more in preventative care, everyone would benefit. Overall spending would drop. Insurance companies would be spending less on claims. Policyholders would enjoy lower premiums and a higher quality of life. More people could afford coverage, increasing profits. It's a win-win situation for everyone.

But it doesn't happen. It's easy to *say* that an insurance company should be bold and forward looking when it comes to preventative care. They often are. There's been an industry push for more preventative care. However, misaligned incentives make it impossible for companies to act alone. Insurance companies struggle to cover forward-looking cost-cutting measures that leave them uncompetitive in the here and now. Doing so would drive away the very policyholders whose health they want to invest in. In order for an insurance company to invest in preventative care, policyholders would have to stay with their insurer when they raise premiums to cover preventative care. In the long run, preventative care would drive premiums down precipitously. In the short term, rates would go up.

In order for preventative care to drive down premiums, policyholders would also have to comply with the care guidelines. Preventative care mostly comes down to healthy lifestyle choices and regular screening. It means eating well, maintaining a proper weight, and exercising. It also means not engaging in unhealthy vices, such as smoking or drinking. In addition, policyholders have to take time out of the day to see doctors and get screenings when they aren't actually ill.

Unfortunately, policyholders are just as bad as insurance companies at weighing short-term pain against long-term gain. They don't want to pay more now for hypothetical future savings. They don't always want to make the lifestyle changes that would keep them from getting sick. They don't want to see a doctor when they don't feel ill.

This is just human nature. Our brains aren't good at running a cost-benefit analysis on future rewards, especially when they are hypothetical. We *may* get sick in the future—but we also might not.

Given the choice, most people would rather save a few dollars on premiums now, stay home and eat that pizza, and hope for the best. Going to the gym might stave off heart disease, but you won't realize the consequences of failing to go until you've already had the heart attack and it's too late. The same goes for mammograms and breast cancer. Not making those annual appointments won't cost you until it does.

Institutions are subject to this same cognitive bias. They are, after all, run by people. For-profit health institutions are incentivized to be efficient, but short-term efficiencies are often placed ahead of long-term needs. Marketplace pressure makes it hard for medical institutions to plan for and allot resources to hypothetical black swan events. We are seeing this now with COVID-19, as of the writing of this book. The American hospital system is facing a shortage of ventilators, hospital beds, oxygen, personal protective equipment, and medical staff because maintaining excess supplies in normal times would make them less solvent. Hospitals didn't invest in preventative resiliency because they had to remain efficient and competitive in the market. When the coronavirus came, it was too late to do anything but scramble.

This kind of bias is a misalignment between current and future needs. It is much less common in industries that don't operate on such extended timescales. Companies in most markets conduct business in the short term. Nike sells you some shoes. The graphic designer makes your logo. The developer builds your enterprise software. The result of cutting corners can be observed right away. The shoe falls apart. The logo looks bad. The software doesn't work. When this happens, the customer walks away or writes a negative review.

Market incentives don't work the same way in the healthcare industry, where prudent investments in preventative care may not show a return for years or decades. Policy could help to better align market incentives with long-term outcomes, but such policy must be crafted carefully to avoid unintended consequences. Much healthcare policy has had the opposite effect of further misaligning incentives.

In an effort to help consumers with preexisting conditions attain affordable coverage, the Affordable Care Act now bars insurers from charging policyholders higher rates based on their current health or lifestyle choices with the lone exception of a smoking surcharge. Plans must treat policyholders the same regardless of health status to be ACA compliant. This protects people with preexisting conditions at the cost of removing incentives to avoid unhealthy lifestyle choices that lead to these conditions in the first place. Insurance companies now have fewer meaningful ways to encourage policyholders to exercise, lose weight, wear sunscreen, or stay on top of regular screenings. Policyholders cannot be financially penalized for most bad lifestyle choices. They cannot be denied coverage when these behaviors lead to a chronic condition.

In many cases, this means that policyholders don't prioritize their health, which leads to worse outcomes and higher overall healthcare costs.

Let's now examine what this looks like on the ground for policyholders.

The Human Cost of an Inefficient Healthcare System

Meet John. John is the everyday policyholder. His story is an amalgam of many real patients. His providers are modeled on real doctors. His insurance companies operate like the ones that I have worked with throughout my career. John may be hypothetical, but his story is very real. Similar stories play out every day all across this country.

John is thirty-five; not old by any measure but not exactly young either. He is a programmer by trade and works in Silicon Valley. His job is great. The pay is good. The benefits are generous. John likes his employer-based health insurance, although he has never had to use it beyond routine care, as he has never been sick. By objective medical markers, John is a reasonably healthy adult. Annual checkups with his primary care doctor always return a clean bill of health. His numbers have always been decent.

However, John doesn't have the best habits. He works long hours in a demanding role. The job is stimulating and rewarding, but the stress can be intense. His project-based work gets extra demanding during crunch time. This leaves John with little energy for healthy activities at the end of the day. He hasn't seen the inside of a gym in he can't remember how long. His diet often involves work doughnuts or bagels for breakfast and employer-provided delivery for lunch. The portions aren't exactly small or healthy.

John knows he should exercise more and eat better. But after a long day debugging code and handling irate clients and bossy project managers, the last thing he wants to do is to hit the gym or spend his few free hours in the evening laboring over a homecooked meal. No, he wants pizza. From Domino's. With breadsticks and maybe a two-liter bottle of soda. Or he wants thick fried noodles from the Chinese takeout joint down the street. But he doesn't want to walk there. He wants Grubhub. He wants Ben & Jerry's delivered by the pint. He wants to plop down in front of the television and veg on Netflix and HBO. He wants to relax.

John tries to practice portion control. Give him some credit. He doesn't eat the *whole* pizza. He saves half of the ice cream for tomorrow night. He's not fat—not *that* fat anyway. He isn't morbidly obese. He's not even technically obese—or at least he wasn't the last time he checked. But he has been putting on weight slowly since college—so slowly he hardly notices. He could probably stand to shed a few pounds—maybe a few dozen really.

John knows he should eat more salads and move around more. But it's hard to fit it into his career-focused lifestyle. He rides a bike to work sometimes—or he did that once. Isn't the bike still in the garage? He's pretty sure it is. Sometimes he thinks about breaking it out and knocking the rust off … maybe tomorrow. Maybe the next day. Maybe. He's pretty tired and pretty stressed. John makes a note to check for statistics on road bike fatalities. He'd better be sure it's safe before he really commits. It's best to be prudent.

The doorbell rings. It's the Grubhub guy with his delivery.

This scene plays out day after day. John tries not to think about it too much. Mostly he is successful—until it all catches up with him at his next physical. John goes in expecting his usual clean bill of health only to get a rude surprise. His lipids are high and so is his blood pressure. Most concerning is his fasting blood sugar. It's in the prediabetic range. His doctor tells him not to panic but also issues a stern warning about needing to make lifestyle changes if he wants to avoid diabetes and heart disease.

"I'm diabetic?" he asks, incredulously.

"Not yet," his doctor warns. "But you will be if you don't start making some changes."

John is shocked. He doesn't *feel* any different. He has always felt healthy. He feels healthy now.

But numbers don't lie. John's abnormal markers put him into a higher-risk category. His doctor is very clear that his current trajectory can lead to serious consequences down the line. He doesn't know exactly when John will manifest symptomatic health problems, or even if he actually will, only that he may someday.

The silver lining: avoiding this fate is entirely within the realm of possibility and is completely under John's control.

His doctor suggests various lifestyle changes. None of them come as a surprise to John. They are the commonsense healthy behaviors John already knew he should be doing. Lose weight. Exercise more. Eat more nutritious foods and fewer empty calories. Cut out added sugar and avoid too much fat. Consume more lean protein and fresh vegetables. No sodas. Less alcohol. Get better sleep. Manage stress in healthy ways.

His doctor hands him some informational pamphlets on metabolic syndrome and a few more on nutrition and exercise before sending him on his way. John leaves pumped to make some big changes to his life. He stops at the grocery on the way home and loads up on salad greens, fresh fish, and low-calorie snacks. His cart is overflowing with broccoli and cauliflower and asparagus. After dinner, he goes online to order a food scale and some free weights. He goes a little overboard with the barbells and dumbbells and kettlebells but shrugs it off. This is worth it, he tells himself. He buys fancy joggers. He also gets yoga pants for stress-relieving exercises.

For a few days, everything is going great. His motivation is high. He's doing all he can. His meals are better. His workouts are regular. Less time gets wasted in front of the TV. His alternating running and yoga sessions are so relaxing he doesn't even miss drinking beer in front of the television. Even the fancy pants are working out great. John is excited. He's pumped about the new life he is creating. He feels proud already. This is a challenge, but he is rising to the occasion.

Then some old college friends come in from out of town. They want to meet up at a local gastropub for burgers and beers. John doesn't protest—after all, he's been so good, he can have one beer and eat something light.

John pays for the first round. He orders his friends IPAs but gets a light beer for himself. For the next round, he switches to water. His friends boo and talk him into another beer. It's on them. John says, "OK, OK, one more." The waitress takes their food order. John weighs the veggie burger against the house salad. His friends order bacon cheeseburgers. John feels jealous and decides, what the hell? His friends are never in town—it's a special occasion! He gets a burger and fries too. Several beers later, they order another appetizer. John goes home feeling stuffed and a little guilty but resolves to do better next week.

But next week at work is *rough*. His boss puts him on another big project while his current one is still active. Now John is juggling two major projects. Worse, the new client turns out to be a real pain in the ass. John's stress levels skyrocket again. The yoga helps, but it doesn't feel like enough.

By Tuesday, he breaks down and gives himself a day off his routine. No workout, no yoga, no homemade salads for dinner. John takes a "me" day. Everyone has a cheat day every once in a while, right? John orders a pepperoni pizza. With Italian sausage. And Canadian bacon. Why not? If you're going to cheat, you might as well cheat right.

Everyone knows where the story goes from here. John slips back into his old habits. The delivery apps he deleted off his phone get downloaded again, one by one. The weights start collecting dust in the corner of the living room. His gym membership lapses. His workout clothes slowly move to the back of his dresser. He's returned to his old routine. The couple of pounds he shed reappear and then some.

By this time, it's been a few weeks since his doctor appointment. John starts rationalizing his behavior again. His numbers weren't *that* bad. He isn't *actually* diabetic. He has time to turn things around. You can eat pizza and be healthy. He will just eat less next time. Of course, he always finds himself reaching for one more slice when he knows he shouldn't. But he's already started, so why stop now?

John has no problem rationalizing his self-destructive behavior in this way. Most people can probably relate. Putting in work in the present for a deferred reward is hard. It goes against human nature. We evolved to react to immediate danger. Our brains reward us with a rush of endorphins for immediate rewards. There's no dopamine kick for turning down that morning doughnut. We also evolved to survive in far more difficult conditions than the modern world presents. Conserving energy is a natural evolutionary trait. Exercising voluntarily and turning down food runs against our biological instincts. We are driven by immediate rewards and default toward the path of least resistance. Our capability for rational thought is the only thing that allows us to engage in these behaviors at all.

But doing so doesn't come easily or naturally. John has to weigh the current pain and sacrifice of healthier choices against the hypothetical dangers that lie far into his future. Working for a deferred reward is hard enough. Making sacrifices to reduce the risk of something that might never happen is even harder. John *might* develop diabetes *years* from now? Distant hypothetical risks are easy to shrug off despite the very serious and grave implications of doing so. Making hard choices now, based on abstract ideas of what might happen far in the future, isn't easy. Some people manage to do it but many fail. John's track record here isn't great.

This is why preventative care is so hard for individuals to sustain. Immediate incentives are misaligned with long-term health. Long-term outcomes matter most—but only in the long run. We cannot feel the full negative consequences of our immediate actions. We *know* about them, thanks to modern medical science, but we cannot really *feel* them. So we put off preventative care. We tell ourselves that we always have tomorrow to get on top of things.

Unfortunately, there's always another tomorrow—at least until the consequences of our bad habits catch up with us. By then, it's too late for preventative care. Reactive care is all we have left—and it's typically more expensive and less effective than prevention. Sure, by then John will wish things were different. But it's too late.

His insurance company may wish things were different, too, when the bills start coming in for John's care.

Fast-forward ten years. John never made the changes his doctor recommended. John is now forty-five and has full-blown diabetes. His doctors have put him on insulin. He has to make regular appointments with his medical team. His health complications from diabetes have to be treated as they arise. His insurance company is on the hook for the care. An ounce of prevention might have paid for a pound of cure, but it's too late for all that. John is already sick.

Of course, the insurance company cannot just eat these costs without raising revenue. They pass the costs on to policyholders in the form of higher premiums, which is why John has been paying so much for health insurance all these years. His high premiums had always made him grumpy when he was young and healthy and didn't need the care. Now that he does, John—and people like him—are driving up the cost for everyone else. John is thankful to have insurance now, expensive or not. Nothing about his premium changes. He pays a little more in coinsurance, but given how much medical care he now needs, this seems like a pretty good deal. He used to feel like his insurance was screwing him—now he is the beneficiary of a system bearing the cost of his bad choices.

John isn't a bad person. He just made flawed decisions in a system that fails to incentivize better ones. John is paying for his poor choices in the form of poor health, but the financial cost was paid up front. His premiums were always high. But because he always had to pay those anyway, they provided no financial incentive to stay healthy. So he didn't. If the long-term price of his choices was paid up front, he might have made better ones.

Healthy people paying for the sick is not a bad thing. Risk mitigation is a valuable service. People who stay healthy their whole lives have no way of knowing that they will until they do. Spreading around risk is the whole point of insurance. The problem isn't the insurance model—it's the misaligned incentives that it creates. The interests of individuals and the collective are not well aligned, which we see reflected in John's choices. The insurance company simply spreads the cost of his unhealthy behaviors across the collective, but this removes the incentive for him to practice healthy behaviors. It's a vicious loop, and its effect can be seen in the spiraling cost of healthcare.

We cannot blame John entirely for this. He is only reacting to counterproductive incentives. The health insurance industry has its own set of counterproductive incentives as well. Imagine a scientific breakthrough in diabetes research that discovers a diabetes vaccine that cuts the incidence of diabetes in half if administered early in life. The catch? Vaccination costs $10,000. Given that the average diabetic patient spends about that much annually on treating the disease, covering the vaccination would seem like a no-brainer for insurance companies.[12] But, from a financial perspective, that may not be the case. Given that John may not stay with his insurance company for life, there is no guarantee that his insurance company

---

12   American Diabetes Association, "Economic Costs of Diabetes in the U.S. in 2017,"
      Diabetes Care 41, no. 5 (2018): 917–28, https://doi.org/10.2337/dci18-0007.

will see a return on that major investment. If John develops diabetes after switching jobs and to another insurance company, he becomes someone else's problem.

Of course, the collective risk doesn't actually change. The insurance industry is still underwriting that collective risk. In the case of diabetes, it is a liability that reached $327 billion in 2017 and has continued growing exponentially ever since.[13] The insurers still pay that cost, but their misaligned incentives mean that individual insurers are doing little to address the situation. Instead, they keep passing the buck back and forth to each other.

This isn't just a thought experiment. When the HPV vaccine was first introduced, many insurers resisted covering it except for the most at-risk groups. Several who could have still benefited were denied coverage, many of whom surely went on to develop HPV-related cancers. Insurers made the choice to not cover them because the vaccine was expensive and there was no guarantee that the cost would be recouped. This is the same mindset that disincentivizes coverage of most preventative care, and although it is bad for the health insurance industry as a whole, it makes financial sense for each individual insurance company.

While misaligned incentives result in underinvestment in preventative care, they also lead to wasteful overspending on reactive care. Insertion of insurers between healthcare buyers and sellers obscures the true cost of care. When John goes to buy food, he knows more or less what he is getting. Grocers include nutritional information on packaging. Menus have calorie counts. The number of calories in a frozen pizza or that side order of fries is knowable. So is the existence of other options. John can find all the local restaurants on the internet.

---

13   Ibid.

The menus are online, and so are reviews. Costs and general quality are discoverable in most retail markets.

This isn't true of the healthcare industry. When John sees his doctor, he generally has no idea what the doctor is billing the insurer. John doesn't even know what he *himself* will pay until a procedure is billed and the claim processed. Patients often have no way of knowing what their care really costs or how much of the bill they will ultimately be responsible for.

When his claims get paid, John couldn't care less what the total bill came to. That's the insurance company's problem. He may be the buyer—but they are the *payer*. They foot the majority of the bill, which gives John little reason to make intelligent choices about treatment. His doctor makes recommendations, and John says OK. The problem is that John's doctor has an incentive to bill as much as possible. Given that John doesn't care, the doctor may bill as much as he can get away with. This isn't necessarily fraudulent care—his doctor just isn't rationing the care appropriately. There's no one making careful decisions about whether a medical service is worth it.

No one is going to ration goods or services without market incentives. When John goes shopping for a television, he considers the cost and value of different options. Maybe he only needs a forty-inch television for a small room. But if someone else is paying, why not get the eighty-inch OLED smart TV? In healthcare, where the insurance company is paying, that is exactly the thought process John goes through when his doctor advises medical tests with unknown or marginal utility. He ultimately pays the bill in the form of higher healthcare premiums, but again, this is disconnected from his own personal behavior.

Costs aside, John also isn't well equipped to discern the quality of medical care. Value, which is price divided by quality, drives most consumer decisions. With most consumer choices, value is easy to identify. John can tell that the big-screen television has a sharp picture. He knows when the hamburger tastes dry or the steak is overcooked. But he is not qualified to evaluate much of his medical care. John knows he likes his family doctor. He's nice and attentive and listens well. But what about the surgeon who will perform his double bypass when paramedics rush John to the ER during a cardiac infarction? There's no time to evaluate the surgeons on staff then.

Even if there were, it wouldn't make a difference. John isn't equipped to evaluate specialists. Suppose he avoided the heart attack by going to see a cardiologist at the first sign of chest pain. John asks his primary care doctor for a referral. John has two choices: Dr. Jones and Dr. Abrams. Unbeknownst to John, Dr. Jones charges the insurance company 30 percent more than Dr. Abrams. This is despite the fact that Dr. Abrams's patients have better health outcomes. They develop fewer complications and infections. How does John make an educated guess with limited information? He can't.

With perfect knowledge, the choice is clear—John should go with Dr. Abrams. He delivers better care at a lower price. But John doesn't have perfect information. John has no way to know what either doctor charges. Even if he did, he wouldn't much care because they are both in his network. John would care that Dr. Abrams's patients have better health outcomes, but with no way of knowing this, John makes the choice basically at random. Dr. Jones has more available appointments and is five minutes closer. John picks him.

The end result: John gets inferior care that the health system pays more for. When John subsequently has to be treated for an infection he developed under Dr. Jones's subpar care, the health system picks up the tab for that too. The bill goes to the insurer who is forced to set premiums higher to cover these kinds of unnecessary costs.

All of this could have been avoided if the person paying for John's care was also shopping for it. But that arrangement just isn't workable in the insurance industry. To compensate, the healthcare sector must find other ways to increase access to information and incentivize better choices. Otherwise, costs will continue to spiral upward.

Dated Information Infrastructure

Access to healthcare information is not just a problem for John. The entire American healthcare system suffers from outdated informational and record-keeping systems. This is particularly problematic given the decentralized nature of our healthcare system.

There are pros and cons to decentralization. It allows for a level of customization, resiliency, and nimbleness not seen in centralized national healthcare systems. However, decentralization can create friction as information moves through and across systems. The American healthcare system involves many actors and institutions communicating across systems that have not been fully automated or integrated. Some of them aren't even fully digitized. When information cannot flow freely through these interlocking systems, decentralization begins to look more like fragmentation.

This is fixable, but it isn't an easy fix. Medical data is highly sensitive information. Failure to keep patient records anonymous and secure can lead to violations of the Health Insurance Portability and Accountability Act (HIPAA). Patient records and data must be handled carefully and securely when traveling through the many different hands that are involved in patient care, billing, and reimbursement. Many actors need to access and work with confidential records. Right now, the fragmented nature of medical record keeping makes that cumbersome and inefficient. Given the sensitive nature of the information, it is also fraught with legal liability. The friction introduced by system fragmentation can undermine any clinical, administrative, or business processes that involve records—which is basically all of them—making the entire system slow and prone to error.

It's also a headache for patients and policyholders. Most of John's current medical records are with his family doctor. When John goes to see a specialist, such as a cardiologist like Dr. Jones, the relevant records—but *only* the relevant records—must be sent to Dr. Jones's office. While John is under the care of Dr. Jones, new records and patient information about his heart condition are created. These records stay at the specialist's office. They are sent out only on an as-needed basis. This is done to protect John's privacy, but in a fractured healthcare system, the onus is ultimately on John to manage his own records.

This isn't easy. Over the course of his life, John has switched doctors many times. He had a family doctor growing up in the suburbs of St. Louis. In college, he saw a string of campus physician assistants in his East Coast college town. Then he found a doctor in San Francisco, but she retired. He is on his second primary care doctor as a working adult. That's to say nothing of his dentists or the dermatologists he sees to track moles.

Each time John jumps doctors, it is like his whole medical history is wiped clean. All of the tests, images, scans, and everything else in his medical record more or less disappear. He can track them down, maybe, for a time. Doctors are usually happy to forward records. But he eventually loses touch with old providers. Many move offices. Some retire. Medical records get lost in the shuffle. His medical history gets wiped clean every few years.

Except, of course, it's not *actually* clean. When it comes to your health, what you don't know can very much kill you. More to the point, what your doctors don't know can kill you. Each new provider has to start more or less from scratch rebuilding John's medical history. They have to run the same tests again.

This only gets worse going forward. Over the course of his lifetime, John will see many more doctors. He will frequent many more pharmacies. He will change jobs every few years and get put on a new employer-based plan with a different insurance company each time. They will all keep their own set of his medical records. They all have some of his records. None have all of them. This complicates both care and coverage. The relevant records have to be assembled each time John interacts with a doctor or his insurance company. Gathering the relevant documents quickly becomes a logistical nightmare. Eventually John can't manage it on his own, but there's no one else to do it except him.

This creates confusion and compromises care. His doctors are making medical decisions based on an incomplete understanding of his history. They can't see how his labs are changing over time. There is no continuity of medical history. No one has a holistic picture of John's health. There are several versions of his medical history in different hands. The whole thing is a mess.

Providers need access to his entire medical history to notice patterns. Risks that should have been flagged may go unnoticed. Health problems might go undiagnosed until it's too late. His insurers process claims without all the relevant context. His pharmacist may not know all the drugs he is taking or about prior allergic reactions. They may have no way of knowing that John is on a potentially dangerous combination of medications.

If John gets into a car accident that incapacitates him, the emergency responders and hospital providers can't ask him about his drug allergies. Unfortunately for John, he is allergic to penicillin. His primary care physician knows that. But the doctors and nurses at the local hospital do not. He's never had to visit the emergency room before. This is the kind of thing that can make what is already an emergency situation much worse.

What John and his medical team need is an easy way to access all relevant records. Currently, the healthcare infrastructure does not make this easy or even possible. Health information systems need to be overhauled and modernized.

Lack of Virtual Care Infrastructure

Information systems aren't the only healthcare systems that could use modernization. Just as records could be better handled over a networked system, so, too, could the actual delivery of care. Telehealth would allow providers to leverage technology to make the delivery of care more efficient and affordable. Many industries already lower costs by conducting business virtually. So far, the healthcare industry is lagging behind in this regard. Patients are accustomed to seeing their doctors in person. Many providers and hospitals are hesitant to make serious investments into telehealth systems that they fear may go unused.

This is why when John goes to see his doctor, he does the same thing that most patients do. He goes in to *see* his doctor. The doctor is at the doctor's office. If John wants to see him, he books an appointment and drives across town to go to the office. John feels comfortable with the arrangement. So do his providers. But the truth is that most routine visits could be performed just as easily online, saving time and resources.

If John could simply see his doctor by setting a ten-minute appointment, he could do checkups while on his lunch break or from the comfort of his home. This would increase compliance and save him time and money. Even if telehealth appointments don't have a lower co-pay, they do have a lower cost for the system, which would inevitably lower overall prices and premiums.

As it is, though, John has to take time off work to go see the doctor. Many patients therefore simply don't bother. Imagine what would have happened to John if he had skipped his annual checkup due to crunch time at work. He wouldn't even know that he is pre-diabetic and would continue on with the same bad habits that would eventually give him diabetes.

Of course, not all care can be given via telehealth. Trips to the emergency room will always involve a face-to-face experience. Some routine exams must be performed in person. But many trips to the doctor could be done virtually. Video chat allows doctors to see and hear a patient just as well as an in-person visit. During the current COVID-19 pandemic, many patients are being diagnosed over video chat. This is being done to take the pressure off an overburdened health system. But there is no reason providers couldn't always see patients in this more efficient way. A quick and cheap video conference can confirm if a patient actually needs to come in. Most trips to the doctor don't require an in-person meeting. Simple questions and requests could be handled by video conference, email, or over a patient portal.

We know this works because it is being done now. Many providers and hospitals do offer telehealth services, but so far, they have not been adopted at scale. Most people visit brick-and-mortar facilities for care. This is seen as the default. When John has a cough, he doesn't go online. He calls in to make an appointment. He makes appointments for new prescriptions, even though this could be done electronically. Doctors are notorious for asking patients to come in for follow-up appointments that could be handled by electronic communication.

This practice isn't just inefficient and needlessly expensive. It's also dangerous. Hospitals and doctors' offices see a lot of sick people each and every day. Hospitals are one of the most dangerous places for sick people to be. Hospital-acquired infections are a real danger. MRSA and other antibiotic-resistant bacteria are very common in hospitals. So are serious respiratory illnesses. People go to the hospital or doctor because they are sick. Limiting your exposure to these settings lowers your risk.

Unfortunately, we still have millions of patients making unnecessary trips to healthcare settings for things that could just as easily have been done from home. This is beginning to change, but we are a long way from the full and best adoption of telecare. We are also only seeing limited at-home care, even though modern technologies could make that a reality. They could also automate the care process even further. In the future, we may see AIs take over for doctors in the delivery of telehealth services.

But none of this will happen unless healthcare providers and institutions adopt new technologies and restructure business practices to best utilize them.

The Healthcare Affordability Crisis

Although misaligned incentives are the source of inefficiencies in the healthcare sector, the ultimate problem is the resulting expense. Many of the problems outlined in this chapter could be forgiven as acceptable operating costs if healthcare remained affordable. That isn't the case. Inefficiencies are driving the cost of healthcare beyond our ability to pay.

The United States has the most advanced healthcare system in the world—but it isn't cheap. Annual nationwide healthcare expenditures now top $3.5 trillion and growing. In recent years, healthcare spending has accounted for 18 percent of the GDP, a share of the economy that will only keep growing. Healthcare spending nationally now exceeds $11,000 per citizen. This is an astounding number considering that the median full-time American worker brings home less than $50,000 per year. The median household income is only $63,000 a year, but the average employer-provided family plan runs more than $20,000 per year in premiums alone.[14]

This is unsustainable over the long term, not only for the healthcare system but also for all of society. The percentage of GDP we spend on healthcare cannot continue to grow unabated. Eventually we won't have money for anything else. This level of healthcare spending puts a major drag on the economy.

The affordability crisis is itself a healthcare crisis. There is a reason the most advanced healthcare system in the world now ranks so poorly against other developed nations—many Americans cannot afford the care they need. The best healthcare in the world means little to those who cannot afford access to it. Too many people cannot afford the best-in-class care on offer in their cities and hometowns.

Rising prices lead to negative outcomes. As premiums, co-pays, and coinsurance are going up, life expectancy and health outcomes are slumping. The United States now trails other countries in various measures of public health despite spending far more on healthcare. We have a higher disease burden, higher mortality rates, and lower life

---

14   Kaiser Family Foundation, "2019 Employer Health Benefits Survey," September 25, 2019, https://www.kff.org/health-costs/report/2019-employer-health-benefits-survey.

expectancy than other developed nations with lower per capita GDPs and lower healthcare spending. We have a greater disease burden and higher hospitalization rates for preventable illnesses. Other countries are seeing better outcomes despite spending less on healthcare.[15]

Some of this is cultural and can be attributed to unhealthy lifestyle choices. The United States has an obesity epidemic. We have an opioid epidemic. Largely preventable chronic conditions, such as cardiovascular disease and diabetes, are on the rise. But these are preventable causes of death and morbidity that a system that emphasizes preventative care would work to lower. Our failure to do so places an ever-greater financial burden on the healthcare system and the country.

What does all this mean for someone like John?

In many ways, John is very lucky. He's young. He doesn't smoke. He is paid well in a thriving industry with solid job security and good benefits. If he were single, his premiums would run him only about $7,500 per year in 2020. However, insuring him and his wife costs double that. If they decide to have kids, they are looking at the $20,000 most Americans pay for an employer-provided healthcare plan. Even with his employer subsidizing the cost of insurance, their family would have to pony up $8,000 in premiums, plus any out-of-pocket expenses. And remember, there's no free lunch. Employer subsidies simply hide the cost of healthcare. It still comes straight out of his total compensation package. The cost of insurance always gets passed back to the policyholder.

Not all Americans are as fortunate as John. Some don't have coverage at all, of course. Others have coverage, just not enough. Although uninsured Americans get more attention in the media,

---

15 Nisha Kurani, Daniel McDermott, and Nicolas Shanosky, "How Does the Quality of the U.S. Healthcare System Compare to Other Countries?" Health System Tracker, August 20, 2020, https://www.healthsystemtracker.org/chart-collection/quality-u-s-healthcare-system-compare-countries.

*under*insured Americans face a significant financial burden as well. Their plans may be expensive and still unlikely to fully cover their liability. In order to realign incentives, insurers have shifted toward cost-sharing models that place more financial responsibility back to the policyholder.

High deductibles and coinsurance do help with the rationing of care, but in such a high-cost market, policyholders may still find themselves bankrupted by a medical crisis. The 2010 Affordable Care Act now caps out-of-pocket maximums at $8,150 for individuals and $16,300 for families (for 2020 marketplace-compliant plans). Premiums do not typically count toward this maximum, which means that a major health catastrophe could cost $40,000 per year for a family with coverage. This is particularly hard on anyone who doesn't qualify for an ACA subsidy. The average American family simply cannot afford that kind of hit without severe financial burden or insolvency. Even with health coverage, many American families are still just one medical crisis away from bankruptcy and losing everything.

This is where we are today. It's only going to get worse from here. The grim truth is that healthcare costs per capita have increased almost every year since 1960. In recent years, that growth has been in the range of 5 to 10 percent, which far exceeds inflation. This isn't any more sustainable for John than it is for the country as a whole. John is well off, but he's not rich. If the cost of healthcare keeps rising, no one is secure. The cost of coverage and care cannot continue to rise faster than inflation without eventually overtaking our ability to pay.

Combine rising healthcare costs with a general affordability crisis, especially in regions with soaring rents such as the Bay Area, and affordable quality healthcare may slip out of the reach of normal Americans sooner than we think. We are already at the precipice. The country is spending too much on healthcare. Individuals are paying more than they can afford.

The rising cost of healthcare is actually a primary driver of the general affordability crisis in the United States. Rising employer-based health insurance costs are reflected in the rising costs of products and services. Health coverage is a labor cost for companies. When General Motors has to pay more for its workers' healthcare coverage, the company must raise the price of its cars. This makes American companies and products less competitive. The lower cost of healthcare in Japan and Germany makes their auto industries—and many other markets—more competitive than their American competitors.

Rising healthcare costs thus become a "tax" on American-made goods. Inefficiencies and disease burden in the healthcare system increase costs. Individuals pass the cost of their bad decisions on to the insurer. The insurer passes it back to them in the form of higher premiums. Companies subsidizing those premiums for their workers then pass the cost on to all consumers. But the buck stops there. We are socializing the cost of healthcare inefficiencies at the national level, but we cannot pass the buck to other sovereign nations. They have their own healthcare systems. If those systems are more competitive, their industries and economies will be at an advantage over ours.

> Exponential technologies provide new hope for increasing efficiency, realigning incentives, improving health infrastructure, and bringing down costs.

In this way, healthcare is having a negative influence on the entire US economy. No industry is left unaffected. Even John, with his high-paying software job, will see his industry becoming less competitive over time. How many of those good software jobs can be outsourced to more competitive countries? We may find out the hard way. His industry and others will have to make difficult decisions about whether to keep paying for healthcare coverage at all.

John cannot fix this problem. No amount of belt-tightening or healthcare rationing on his part will solve a systemic problem. The most expensive healthcare is inelastic. If you have cancer and need chemotherapy, you need chemotherapy. When you get in a major accident, you need to use the emergency room. He *could*, in theory, be better about managing his own health and taking preventative care seriously, but he cannot force the millions of other Americans to do the same.

His employer or insurance company cannot fix the problem either. They operate under market pressures to stay competitive in the short term at the expense of long-term efficiency. This is a societal problem. The healthcare industry will have to collaborate with all relevant actors to realign incentives throughout the system so that we are not all pulling in different directions. Institutions and enterprises must work to reform and restructure the whole healthcare system to be more efficient.

This is a major undertaking. It won't be easy. However, we do have new tools at our disposal to start righting the misaligned incentives that have the whole industry working against itself. Exponential technologies provide new hope for increasing efficiency, realigning incentives, improving health infrastructure, and bringing down costs.

This book focuses on two such technologies—artificial intelligence and blockchain—which are poised to help us remake the healthcare industry and much of society. But we have to be willing and able to deploy them to solve the unique business problems that the healthcare sector faces.

First, let's look at the present and future potential of artificial intelligence.

CHAPTER 4

AI EXPLAINED

THE PUBLIC IMAGINATION of artificial intelligence has been heavily shaped by science fiction. The average person conceives of artificial intelligence as digital consciousness. The term *AI* evokes images of Data on *Star Trek* or Arnold Schwarzenegger's similarly robotic performance in the *Terminator* franchise. We think of loveable robots like WALL-E from the eponymous film or Johnny 5 in *Short Circuit*. C3PO from *Star Wars*. Samantha in Spike Jonze's *Her*. David in Spielberg's *A.I.* The holographic Rimmer in the British comedy *Red Dwarf*. It's almost harder to find science fiction narratives that don't feature an AI character.

Actual scientists and technologists refer to this kind of humanlike AI as general artificial intelligence. General AI attempts to mimic the kind of abstract thought and typical problem-solving skills seen in humans. I say *attempts* because there is currently no existing general AI system even approaching the sophistication of the human brain. The technology is nowhere close to creating a system capable of abstract thought or general intelligence.

It is not for lack of trying. Some of the earliest research into AI was on general intelligence. Engineers have long been trying to create AI systems that can pass the Turing test, a concept derived from scientist Alan Turing's 1950 research paper, "Computing Machinery and Intelligence." AI was becoming a matter of serious scientific discussion at the time. There was an ongoing debate about whether AI systems could ever really "think" in the ways that humans do. Writing for the journal *Mind*, Turing argued that the question was unscientific because there was—and still is—not any consensus on what it even means to think. Without clear parameters to test, a hypothesis is unfalsifiable and therefore unscientific.

Turing redirected scientific inquiry toward the ability of AI systems to *appear* human, which could be measured and quantified. Turing devised a simple experiment to test how well human subjects can discern AI chatbots from real humans. Thus, the Turing test was born. Scientists and engineers immediately started trying to create AI systems that could fool observers into believing they were humans.

The Turing test is not a single test—it is a concept. Different Turing tests can involve different controls and measures. How long must contact be sustained? Does the observer know that they might be chatting with a bot? The earliest AI systems could only fool humans within narrow constraints for a limited time. However, as AI technology advanced, the chatbots became better and better at fooling humans under a variety of conditions.

Although the Turing test has been a great vehicle for spurring research into artificial intelligence, the test is not a good measure of general intelligence. In fact, by definition, the Turing test is no measure of general intelligence at all. That was the whole point. Turing thought testing for general intelligence directly was a dead end. The

Turing test simply measures our perception of an AI's possession of humanlike general intelligence.

For some, the insinuation here is that there is no functional difference between general intelligence and its appearance. This leads us to a philosophical inquiry beyond the scope of this book. In practice, the point is moot. There are many AI systems that have "passed" Turing tests—but not one of them actually exhibits general intelligence approaching anything near the human capability for broad abstract thinking.

The scientists developing these AI systems aren't even trying to create systems that function like the human brain. This isn't possible with modern technology. Instead, the developers are teaching for the test, in much the same way that IBM taught Watson to play *Jeopardy!* This can produce impressive results, but it is not general intelligence. It is not how the human brain works.

General AI versus Applied AI

The human brain is amazingly flexible. It can solve an unimaginably broad number of problems. Humans are capable of language, math, art, logic, athletics, and much more. Any one of these disciplines contains unlimited branches and subbranches and sub-subbranches all the way to infinity. Humans can excel at any number of them with enough practice. No current AI system has such a wide range of capabilities.

In fact, AIs cannot even perform simple actions that rely on decisions we take for granted each day. They cannot navigate social situations. They cannot drive a car safely on their own. They still struggle to control robots that need to navigate unknown terrain that humans could nimbly traverse with ease. These kinds of tasks often require contextual analysis. Fortunately, brains are very good at understanding context. Evolutionary pressure trained us to master context over thousands, even millions, of years.

Current AI systems, for all their computational capability, do not have this ability. Artificial intelligence relies on decision trees. Adding variables adds forks. We can always define new variables, but understanding complex context requires the ability to handle an infinite number of possible variables that all interact with one another.

Human brains do this well. We can see a barking dog and instantly determine the threat level. Sufficiently advanced AI can easily recognize a dog and even analyze the pitch and cadence of barking or growling. However, the system may struggle to factor in the breed, or if this dog has been encountered before, or the location, or whether the dog is capable of jumping the fence between them, or a million other possibilities. Humans are able to connect the dots and judge context in a way that is just not obvious to machines. This allows us to solve problems creatively and to think abstractly in a way that machines cannot.

Current AI systems can handle a tremendous amount of complexity, but they still cannot handle limitless context. However, when AI developers introduce constraints on context, AI systems can make decisions far more rapidly and accurately than humans. Within constrained applications, AI systems are much better at processing and synthesizing information than the human brain. Computational power allows them to process information at lightning speed. They can crunch massive amounts of data and perform the same operations repeatedly much faster than humans. This is why AIs can beat chess masters and trivia champions while struggling with activities that require the most rudimentary abstract or creative thought. Chess is a complicated game, but it can be easily broken down into moves and optimal strategies. There are strict constraints on what you or an opponent can do on each turn.

The strengths and limitations of the current generation of artificial intelligence make it most applicable for solving fit-for-purpose business problems. AI systems are designed to handle a very specific task or type of task while operating within imposed contextual constraints. They are particularly useful when applied to highly repetitive computational tasks that would require humans much effort to figure out by hand. They can organize a library of texts or music almost instantaneously. They can sort, analyze, correlate, and process records. They can crunch numbers to encrypt or unencrypt massive amounts of data. And they can do these things in a fraction of the time that it takes humans.

These fit-for-use systems and tools are known as applied AI. They lack the broad general problem-solving of general intelligence but are powerful problem-solving tools when operating within constraints. This makes them perfect companions to real live humans who are able to design and operate them with contextual knowledge of how they will be applied while relying on their computational power to synthesize data and make decisions or recommendations accurately.

There are scientists working on general artificial intelligence, but we are in the early stages. Personable robots and conscious mainframes are going to have to keep waiting. The technology is simply not there yet. We don't have the processing power. We don't know how to build algorithms that can mimic human thought processes. We don't fully understand how the human brain works or how to define consciousness. Getting an AI to think like we do is just not currently possible. We are still decades away from AIs with the sophisticated general intelligence depicted in science fiction. We are making advances. I truly believe we will get there someday—but maybe not in my lifetime.

However, we do have artificial intelligence systems that we can deploy to solve current problems that we could not do nearly as efficiently, if at all, without them.

Applications of AI Defined

There are four primary applications of applied AI in industry:

- Automation

- Insights

- Personalization

- Sensing

These are all fit-for-purpose applications that are perfect for applied AI. They are narrowly focused applications that operate well within constraints. Constraints are not a problem when you are operating with a narrow focus. Most business processes are limited in scope and occur within a specific context. They don't require much contextual analysis. AIs are therefore able to handle these processes well with low levels of error or inaccuracy. The current generation of AI has no difficulty solving many problems in these areas.

Let's look at these applications individually and see how they function synergistically to create efficiencies and add value.

Automation

Industry has long been automating business processes. Automation saves countless man hours and resources, as computers can process data in a fraction of the time it takes humans. There are all kinds of use cases. Easier and more accurate data input. Automated paperwork.

Self-service systems that allow for effortless customer service and data collection. Automated quality controls.

Some of these processes require decision-making. This is where AI comes into play. The system needs to be able to synthesize data that will influence how it acts. These are not complex executive decisions that will decide the fate of companies. (That kind of higher-level decision-making requires general intelligence.) These are mostly tasks that require simple, repetitive decision-making. Should this form be forwarded for further review? Should this claim be denied? Should we accept this record into the system or request more information? Should we order more or less of this product for inventory? Does selling this call align with predetermined investment strategy? These aren't the sexiest, highest-level decisions being made in the business world, but they do comprise the bulk of the analytical and administrative work that many businesses do each day.

Applied AI excels at making these kinds of decisions. Most business processes involve a structured set of inputs. Decisions are made based on defined policies and guidelines. The variables in the equation are well known and operate within a narrow context. Algorithms can make these decisions rapidly and accurately.

These algorithms allow much of the workload that companies do to be offloaded onto automated systems. Banks use AI to automate the loan process. Relevant financial records are collected automatically, validated, and analyzed—the system can give a recommendation on the loan before a lender ever sees the application. Bad loans that would just be rejected can be identified and denied without human eyes ever seeing the application.

This may sound frightening and impersonal, but these systems actually assess loans more accurately and fairly than humans. They look at a borrower's credit history, credibility, liabilities, and other factors and make an impartial decision. Quite often, these individual metrics are also calculated by AI. For example, credit scores are calculated automatically. These metrics then get synthesized with other metrics to reach a decision on the risk of a loan. Much of the potential for human error and bias is removed completely.

Insurance companies have mostly automated the claims process. The computer system is fed data on the claim and taught to make decisions in accordance with how humans have historically decided on similar cases. Edge cases get flagged and sent for human review. These systems have a first-round accuracy on par with human review. Their use also frees up claims adjusters to focus on reviewing edge cases that need careful consideration. The efficiency of the AI system at automating the process allows humans to focus attention where it is needed.

> As long as decision-making remains within necessary contextual constraints, computers often outperform humans at many business processes. There is simply no reason not to automate as many as possible.

AI systems can return quite complex results. Some claims are simply denied. Others may need to be adjusted, which the systems can do automatically. Multivariable analysis goes into these decisions. The system can retrieve information from a range of sources and synthesize it all at once. Documents can be pulled and read automatically. Automobile insurers may pull police reports, insurance guidelines, personal data, and other information from various sources to make a specific recommendation beyond just "accepted" or "denied." These systems really can perform 95 percent or more of the job of a claims adjuster of yore.

Sometimes they do an even better job than their human counter-parts. Insurance companies instruct AI systems to approve claims that, while illegitimate, are too costly to fight in court. Small claims may not be worth the cost of litigation. Some claims can be simply adjusted down low enough that they are unlikely to trigger litigation. These are decisions that human adjusters might struggle to make. They rely on multivariate complex analysis to arrive at unintuitive decisions. Claims adjusters might be biased toward accuracy over financial prudence. The AI isn't burdened with such preconceptions and can be trained to make decisions on any metrics without the risk of impassioned bias.

The returns for automating menial decisions are so substantial that this kind of AI is widely applied. Automating away repetitive tasks frees up resources and allows workers to focus on other things. The AI system typically performs automated tasks faster and more accurately. As long as decision-making remains within necessary con-textual constraints, computers often outperform humans at many business processes. There is simply no reason not to automate as many as possible.

Insights

The data produced by automated business processes holds valuable insights. These insights can be about almost anything. They may allow businesses to better understand the context in which they are operating. They may reveal something about a business process or stakeholder. They may unlock untapped opportunities. These insights can even allow companies to make predictions about the future.

Providing examples of insights is almost unnecessary. All business data contains insights. An insight is simply a pattern that reveals something about the data. All data produces insights. Insights can be used to model markets, marketing trends, pandemic outbreaks, climate change, and countless other things. There really is no limit to what we can learn from data. There's no limit to how we can apply those insights.

The data analysis that leads to insights is mostly automated. Given the volume of data that companies now work with, there is no other way to analyze and handle it. Automated AI systems are essential to collecting, organizing, and analyzing data. We need these systems to find the signal in the noise.

Researchers and analysts have been using algorithms to glean insights from data for a very long time. This isn't necessarily artificial intelligence. Setting variables, defining operations, and getting a return value are often just crunching data. Artificial intelligence comes into play in two ways. The first is in the actual pattern recognition that generates insights. The second is in decision-making about what to do with the data and insights. AI systems don't just mindlessly execute code. They look for patterns and make decisions based on training and defined guidelines.

The exponential power of these systems can be seen in machine learning. AI systems are able to return insights about insights into their own operation. This allows them to improve their data sets and algorithms. This is a kind of digitized evolution that builds on itself through multiple iterations. This lets AI systems draw insights that the designers may not have ever considered or been looking for.

Personalization

Insights have many applications, but one that is rapidly transforming whole industries is personalization of user engagement. The insights drawn from consumers, users, and other stakeholders can be used to improve their experiences by engaging with them in a personalized manner.

In our highly connected world, companies know a lot about the stakeholders they serve. This is particularly true of tech companies and their users. Social media and online commerce companies know our places of residence, likes, dislikes, online browsing and shopping habits, travel interests, politics, daily movements, and much more. Companies use this data internally to build individualized profiles of their customers. The data can also be traded and shared between companies to create profiles on potential customers.

There are many ways that companies can use data to personalize the experience for the stakeholders they serve. Retail companies have their websites feature inventory designed to appeal to the specific viewer. Coupons and deals are tailored to the consumer. Online ads are microtargeted at individuals based on insights into their specific consumer behavior. Social media platforms also customize newsfeeds to increase user engagement. Search engines provide results that are tailored to the user's location, demographics, search habits, online shopping history, and other data. Some companies even alter the aesthetics of their websites to be more appealing to the specific person viewing them.

This kind of insight-driven personalization relies on applied AI. The system takes in data and draws insights that allow it to make *decisions* about how to personalize the user experience for each user. This can be as straightforward as offering consumers products, services, or experiences similar to what they have bought or interacted with previously. A Facebook user who clicks on *Economist* articles might also be served up content from the *Financial Times* or the *Wall Street Journal*. Someone who "likes" the *Harper's* page might start seeing suggested articles from the *Atlantic* or the *New Yorker*.

These may seem like obvious associations. They are. Anyone could guess that a reader of *Sports Illustrated* might be interested in *Grantland*. However, Facebook doesn't have to guess—it doesn't even have to think about it. The AI does the thinking for Facebook, which allows personalization to be efficiently applied at scale. Algorithms select "handpicked" content based on user profiles. The AI systems can draw insights without engineers having to program the algorithm specifically.

This is helpful because not all insights are so straightforward. The massive amounts of data that these companies hold can offer surprising insights that no one could have imagined. Comparing similar users allows AI to make associations that people could not. The online dating website OkCupid is known for its data-driven matching system. The company famously discovered that the trait most associated with whether someone will have sex on a first date is the fact that they like beer. This is the kind of insight few would have imagined—similar strange associations are hidden in consumer data.

AI-driven personalization has been revolutionary for the marketing industry. The microtargeting that companies do online is very different from how companies used to target consumers. In the past, companies and marketers targeted broad demographics, not individuals. Most television advertisers focus on the eighteen-to-forty-nine demographic when looking at Nielsen ratings because these are the most active consumers for most consumer products. Of course, certain companies target other demographics. There is a reason that the commercials during daytime television feature so many workman's compensation attorneys, trade schools, and infomercials pedaling goods for stay-at-home moms. This is a very crude way of targeting advertising based on demographics.

The rise of the internet turned that strategy on its head. Online advertising targets individuals instead of demographics, allowing for a more personalized consumer experience. Marketers don't box people into giant demographics as much. They examine individual preferences and market accordingly. This wasn't possible before the rise of the internet, smartphones, social media, and online commerce.

> The healthcare industry produces massive amounts of data that could be harvested, organized, and analyzed to personalize all aspects of care.

Ads and product recommendations are not the only thing that can be customized to individuals. The healthcare industry could use AI-driven insights to personalize patient experience and medical therapies. Until recently, the healthcare industry was still stuck in the Nielsen era. Treatments were generic and almost entirely uniform. Sometimes clinicians would adjust therapies to account for broad demographic differences. That was about the only level of therapy customization outside of rehabilitation and physical therapy.

This is finally beginning to change. Patient data is now starting to be used to customize patient care to the individual. This is not demographic stereotyping. AI tools allow clinicians to consider several variables specific to the patient, including their personal lifestyle choices and medical history. This information can be compared against user data in a database to design custom-tailored therapy plans.

This is happening now—but it could be happening much more. The healthcare industry produces massive amounts of data that could be harvested, organized, and analyzed to personalize all aspects of care.

We could also be tracking health, medical, and drug data with more granularity to provide even better insights. Trials typically report efficacy and the percentage of side effects. Holistic data on *who* had side effects doesn't always get recorded. The better we track and record this information, the better our data and the more powerful our insights. These insights have the potential to create a healthcare system personalized to produce the best possible patient outcomes.

Sensing

Sensing is another major business application of AI. Sensing produces a kind of insight that deserves special mention due to its revolutionary potential.

Insights predate AI and even computing. Insights are simply pattern observation. They allow us to better understand the past and present. They also allow us to make projections into the future. The exponential explosion in data and computing power now allows us to better recognize patterns as they are forming and predict how they will develop, which in turn allows us to actually sense trends as they form and develop. Essentially, insights now let us predict the future with increasing accuracy.

This ability has huge business applications. Detecting and capitalizing on trends as they emerge allows for first-mover advantage. Locating the epicenter and direction of a new trend as it forms is hugely valuable. Imagine being able to detect major fashion trends as they emerge. You could be the first to start manufacturing product to meet the growing demand. Imagine being able to detect a hot new musical scene or genre before the bands are signed to major labels. You could swoop in and sign them first.

Viral trends only seem to come out of nowhere due to the nature of exponential growth. We notice them only when they explode. Detecting the epicenter first requires automated monitoring and AI tools set to search for new patterns in an industry or market. The questions for companies and AI teams are: How close can we get to the epicenter? How quickly can we identify the patient zero of a new trend?

With current AI technology, the answers are often *pretty darn close, pretty darn quickly.*

The data exists. The trillion-sensor economy is upon us—and it won't stop here. How long before the next trillion sensors are deployed? The data will be there. We have the computing power for AI analytics that can find patterns. Machine learning allows us to build algorithms that can teach themselves to find new insights. Everything is in place to spot new trends and even predict them before they emerge.

The economic value of prognostic sensing is immense. Consider the sudden rise and success of TikTok. ByteDance, the Beijing-based parent company that developed the app, hit it big filling an emerging niche for teens who could use a platform for recording and sharing short videos. This wasn't a novel idea. Other sites, such as Vine, had done the same thing. Short videos could also be shared as Facebook "stories." But TikTok managed to notice the trend among teens, built a dedicated app, and marketed it to the emerging user base as the trend developed. ByteDance was the first to pick up on all these teenagers posting silly dance videos. They moved first—TikTok is now worth billions of dollars and is changing the social media landscape.

This was a trend worth billions of dollars to the company that moved first successfully. The question now is—what's the next one? And how can *you* sense it first?

The answer will involve leveraging AI sensing technologies. There is a whole emerging science and industry around trend sensing. In the future, no one will just happen along such a niche. The first-mover advantage will almost always go to those actively sensing trends.

Anyone waiting on the sidelines to catch trends as they go viral is going to be left in the dust.

This isn't some distant future. It's happening right now. There are many AI firms specializing in trend sensing for their clients. Spotify uses sensing to identify trending bands and genres. They monitor music communities to sense new trends in music, be it a new artist, a new sound, or a new scene. They even work to predict future hit songs and groups that are likely to become global phenomenons. Automated systems are always analyzing the data in their systems to find that which is about to go viral.

Sensing technology isn't limited to tech companies. Legacy industries apply AI in the same way. In the auto industry, engineers are using sensing to predict the designs that will be popular the following year. Traditionally, automobile designers would develop a concept by hand, build a prototype, and undergo several rounds of feedback and redesign before starting to manufacture. This process takes a lot of time and money. If the design is a dud, billions can be wasted.

In order to mitigate risk, car companies now look for insights into developing trends in consumer automobile culture and habits. Car companies are looking to fill tomorrow's consumer demands today—and they are succeeding. AI shines a light onto developing trends. Designers no longer have to operate in the dark. They have insights into what designs are most likely to succeed. This lowers the risk of going to market in an industry with very high overhead costs.

Every industry has trends to follow. They can involve ideas, designs, concepts, products, fashion, the arts and entertainment, or anything else. In the healthcare industry, trends can include lifestyle habits (good or bad) that can have disease burdens that need to be addressed or planned for. There are also epidemiological trends to look out for, such as viral outbreaks, trends in chronic illness, or the spread of antibiotic resistance among bacteria. The ability to predict the immediate future can reveal all kinds of market opportunities and potential efficiency gains.

The Synergistic Power of Applied AI

The applications of AI just outlined are not discrete business strategies. They can—and should—be deployed synergistically in an AI-forward business model. These applications build on each other. *Automation* is necessary to collect, parse, and draw *insights* from large volumes of data. These insights can be used to *personalize* the product or service, which in turn helps collect more and better data. Better data allows for better *sensing* and even better insights. The synergistic nature of these applications of AI produces compounding returns that drive the exponential growth of the technology.

Consider how these applications of AI are used in concert by location navigation services. These services are now big business. Garmin and MapQuest paved the way by making GPS navigation a mainstream consumer technology. Google Maps and Apple Maps then leveraged their ecosystems and the rise of the smartphone to dominate the industry. Ride-sharing/ride-hailing apps, such as Uber and Lyft, also rely on the same technology to match drivers with riders and to route rides.

These services, which all function in basically the same fundamental manner, leverage AI in various ways. They are predominately online platforms, allowing most of their business processes to be *automated*. They collect user GPS location, as well as speed and direction, to offer turn-by-turn directions in real time. Artificial intelligence is used to make automated decisions about which route to take. The best route at any given time depends on road conditions and traffic. The AI system considers all these variables and decides on the best route automatically.

As these services matured, they began to use AI to actively manage traffic. They route cars in a manner that reduces overall traffic. They don't just avoid traffic jams—they work to actively prevent or reduce backups. By acting like a centralized "brain," these AI systems are running our individualized road transport system more like a centrally planned public transport system.

This is done by collecting *insights*, which can get very complex. They do far more than just count the cars on the road. The AI systems factor in the time of day, weather, holidays, large events such as sports games, and other variables that affect traffic. They use insights and, in the case of apps like Waze, real-time user feedback to detect traffic problems as they arise. They are constantly searching for patterns and new insights. These insights allow them to *sense* problems as they emerge, thus predicting future conditions.

Sensing is also used to predict trends in the industry. Waze caught on by allowing users to input feedback, a novel feature that proved to be popular. This opened up the possibility of new features, such as tracking gasoline prices and routing users to the stations with the best deals. There are other monetizable trends waiting to form and be discovered.

Modern navigation apps also use their data to *personalize* the user experience. They suggest places of interest along the route and may even choose between comparable routes based on possible interests or user preference. This personalized functionality can be a source for monetization through hyperlocal advertising, which is how Waze brings in revenue.

These companies have built fortunes and improved the world for billions of people. They did so by leveraging exponential technology and riding the development curve. They made use of the latest trends in technology to improve and expand on their services. These services started as a way to get real-time directions on the move. Punch in an address and the system would tell you where to go.

The next level of maturity involved layering in the new insights that allowed for traffic management. The algorithms became ever more sophisticated as more data was overlaid to account for weather and other conditions. The apps went from not only telling people *how* to get somewhere but also *when*. The driver could now plan their trip or commute around the best time. More and more of the routing is being handed over to the AI system. In the case of ride-sharing apps, the customer doesn't have to deal with any of this. The service handles everything.

The history of these services has been to require less user input. First, the driver could let the AI pick the route. Then the driver could let the system select the best time. The next big step in this process will be removing the driver from the equation entirely. Uber and other companies are developing autonomous vehicles that will allow AI to actually do the driving. GPS and onboard sensors are used to determine the location of vehicles on the road, not only for routing purposes but also for fully autonomous navigation.

We can draw a line from the early GPS navigation tools to the autonomous cars of tomorrow—and that line tracks an exploding exponential curve. These services rely on AI that is getting better and better. The companies offering them utilize leveraged data to create insights and provide a service. Those insights and the service offer more data and better technology, which in turn provides even more data and better insights. Couple this with exponential progress seen in other technologies, such as smartphones, and we have a curve going parabolic so fast that it's giving rise to whole new industries built atop the tech. The first ride-sharing apps were founded barely a decade ago—now they have transformed the way we do transit.

The exponential growth isn't even close to slowing either. Autonomous vehicles will do away with commercial drivers. Truckers will become a thing of the past as the technology is deployed to move freight. The whole system will run with minimal human oversight by engineers. Eventually AI systems will do much of that oversight as well. We could one day have the whole national transportation system automated and running itself. The highway system will be run the way the airlines are managed now. Pilots and air traffic controllers mostly just monitor systems and provide emergency intervention. Piloting and management of air traffic are nearly entirely automated. Eventually ground transport will be just as fully automated. We may soon no longer lose forty thousand people a year to automobile accidents, as automated cars are likely to be just as safe as automated air travel.

We can likely achieve the same level of AI-powered efficiency and improvement in all industries, including healthcare. Someday care will be almost entirely done by machines. Doctors will be virtual. Robotic surgeons will perform flawless operations every time. The system will be free of inefficiencies, friction, and suboptimal outcomes. AI will one day get us there.

In the meantime, we have to walk the path and climb the curve. The healthcare industry needs to take the same all-in approach to AI that tech companies are doing. We will see rewards every step of the way as we employ the latest applied AI wherever we can.

In the next chapter, we'll look at ways that the healthcare industry of today can do just that. But first, let's take a minute to consider the ethical considerations surrounding artificial technology.

CHAPTER 5

ETHICS AND AI

EXPONENTIAL TECHNOLOGIES WITH the potential to drastically change the world are powerful. With that power, as the adage goes, comes great responsibility. We must consider the serious ethical dilemmas and concerns that surround new technologies before and as we deploy them.

The popular imagination of AI being so focused on general artificial intelligence as it appears in science fiction causes the average person to be most worried about the peril of runaway machine consciousness commanding armies of human-killing robots. Whether such concerns are realistic in the future is irrelevant for now. General AI is nowhere near the level of sophistication that should cause us to have such worries.

However, there are less dramatic ethical concerns with the use of the applied AI systems that we have now. The two big ones that the industry is contending with today are issues regarding *safety* and *bias*. They may not be as dramatic as robotic armies of superintelligent, man-hunting androids, but they deserve our attention all the same.

Safety

When automating large swaths of the economy, including the function of actual machines in the real world, we must consider the safety implications. We are trusting AI systems with our very lives. They fly our commercial airliners. They control our urban infrastructure and utilities. Smart-monitoring camera systems are on our streets. Security robots are being deployed in malls and other public spaces. AIs are helping perform our surgeries and fight our wars. They manage safety systems for construction sites, hospitals, public transport, nuclear power plants, and everything else.

Thankfully, automated AI systems have an excellent safety record. That's precisely why we use them. They reduce human error. Autonomous commercial aircraft have an incredible safety record. Your odds of being struck by lightning or attacked by a shark are many times higher than the odds of dying in a plane crash, especially in an American commercial airliner. This is largely thanks to the AI systems that control the plane.

Yet many people who happily commute an hour to work despite the one in one hundred lifetime odds of dying in a car crash are still scared of getting on planes. Humans are not always good at estimating real risk. This is partially a function of the primal fear that comes with knowing you are tens of thousands of feet in the sky. However, much of the risk is from not being in control of the plane. Many people would be absolutely horrified if they knew that the plane was piloted entirely by computer systems, even though it is these very systems that keep them so safe. People fear new and unfamiliar technologies, especially when they feel like it removes their sense of control, no matter how impeccable its actual safety record.

This kind of irrational fear may hold back development and adoption of autonomous cars. In 2018, Uber temporarily shuttered its self-driving car testing program after an autonomous vehicle struck and killed a pedestrian in Tempe, Arizona. The vehicle's AI system didn't recognize the victim crossing the road at night, because she was jaywalking. The AI wasn't expecting to see a pedestrian appear suddenly outside of a crosswalk.

Tragically, the safety driver inside the car was streaming *The Voice* on her phone rather than watching the road when the accident happened. Had the driver been paying attention, she would have been able to recognize the victim as a pedestrian and perhaps would have swerved or stopped in time. Unfortunately, she wasn't. The victim died.

The failure of the AI system in this situation was a limitation of applied AI. Controlling airplanes in wide-open skies turns out to be much easier than terrestrial vehicles operating on busy roadways with people, wildlife, and other cars darting about. Highways full of self-driving cars would likely lower car collisions. The cars would all be tracked and decked out with sensors. They would be less likely to collide than humans are when weaving in and out of traffic at rush hour. However, enter the unknown of human-controlled vehicles and other hazards, and suddenly our AI systems have a harder time keeping us safe from unexpected dangers.

In the case of the Uber crash, the AI system wasn't able to recognize the pedestrian out of the expected context. Although humans are so good at contextual analysis that we can literally zone out while driving and then snap back into focus in a split second when a deer bolts onto the highway, the current generation of applied AI systems still struggles with the contextual analysis needed to make autonomous cars completely safe on the road. AI systems are great at avoiding

> **At the end of the day, human error is the cause of more accidents and deaths than any currently operating AI systems. Well-implemented automation has more potential to help keep us safe than it does to harm us.**

known dangers, such as other cars, but quite limited when it comes to certain unexpected situations that require contextual analysis to react properly.

Until we have general AI systems that can handle complex contextual analysis on the fly, designers of autonomous cars are going to have to build their algorithms carefully. They will also have to make hard decisions about whether the technology is really ready for prime time yet.

As a society, we must make clear-eyed decisions about risk tolerability and acceptable levels of collateral damage. We must learn to be more rational about risk assessment and identify our own biases. Expecting AI systems to be perfect is not reasonable. While the AI system in Uber's autonomous vehicle failed and resulted in a death, so, too, did the human who was supposed to be paying attention. At the end of the day, human error is the cause of more accidents and deaths than any currently operating AI systems. Well-implemented automation has more potential to help keep us safe than it does to harm us.

This isn't to minimize the risk, though. Fully automating technologies that present risk to property and life do create special concerns. AI systems develop efficiencies by operating at great scale. Any errors in judgment made by the AI system can scale up across the whole system. When a human driver makes a mistake, the car gets into a crash. When the AI in a commercially deployed autonomous vehicle makes an error of judgment, that same error gets repeated over and over again across the entire fleet.

Design choices also create ethical dilemmas that are not always considered until a system is automated. Design choices are made up front and therefore must be well thought out. These ethical dilemmas are often easier to overlook when they are made by instinct on the fly. This is illustrated well in a classic conundrum facing the designers of autonomous cars. What to do in the event of an unavoidable crash? Should autonomous vehicles prioritize the safety of the driver or an unknown pedestrian? Is it better to strike the pedestrian or risk swerving into a ditch or another car? Is an elderly couple more expendable than the parents of two young kids? Should it matter if the pedestrian is jaywalking instead of using a crosswalk? If so, how much?

There are no simple answers to these questions. The issues are subjective moral quandaries well beyond the scope of this book. However, they are design choices that developers and planners must make up front when designing AI systems. These design choices have serious ethical implications.

They are also the kinds of decisions that are part and parcel of the healthcare sector. We will have to keep such issues top of mind when automating an industry that deals with life-or-death situations on a daily basis. Design errors can cause AIs to replicate mistakes across the entire system. When we are talking about hospital and health systems, the stakes could not be higher.

Artificial Bias

AI systems are often trained to make decisions as humans would. This allows business processes that were once carried out by people to be offloaded onto computer systems. A risk of AIs mimicking human decision-making is that they may do so a little *too* well—mimicking human judgment, warts and all. Humans are not always rational or optimal decision makers. We are plagued by biases. Unless designers take steps to ensure that the algorithm recognizes and corrects for bias, the system may repeat our own bad habits. AI systems can only make decisions based on the data and guidelines we provide.

This can have severe adverse consequence when bias is replicated at scale. There have been several cases in which AI systems have demonstrated racial bias. In 2016, Microsoft launched Tay, a Twitter chatbot that was intended to interact with teens online in order to learn the casual conversational language of young people. By chatting with teens on Twitter, Tay was supposed to learn to incorporate their lingo, diction, and cultural references in order to speak more naturally.

Unfortunately, Microsoft had to take Tay down when the system started picking up bigoted language. This was sometimes the work of trolls intentionally training Tay to tweet offensive slurs and jokes. However, Tay was also likely picking up bigoted and offensive language naturally as well. The bot was designed to incorporate its interactions into its own language and knowledge. If there are racist and misogynistic elements on Twitter—and there are—it was just a matter of time before they worked their way into Tay's lexicon.

Tay was only making decisions about how to speak. Other AI systems make decisions that can have real-world consequences. There have been numerous examples of systems engaging in racial bias that range from the comical to the pernicious. AI-powered systems, such as sinks and soap dispensers, have sometimes failed to activate for people of color. Some Xbox owners reported that Microsoft's Kinect camera would only detect white people under certain lighting. These problems were the result of infrared cameras that had, apparently, only been tested on white people. In a similar and particularly egregious

AI-caused gaffe, the auto tag feature of Google Photos started tagging black people as animals because it hadn't been properly trained to recognize different skin pigments.

These problems, each one its own PR nightmare, could have been avoided through better testing of the data going into the system. AI systems are only as good as the data they get. Bad data means bad insights. Garbage in, garbage out. Because the people conducting the testing were presumably mostly white, they missed how their systems interacted with data from minority groups. This was predominately a data collection problem resulting from a lack of minority representation in the groups that build these systems.

Other kinds of human bias get ported into AI systems more directly. This can have serious implications. If minorities are denied home loans more than white people, AI systems performing loan analysis may correlate race (or some proxy of race) with loan risk. Many police departments now employ predictive policing technologies. These are black box AI technologies that analyze crime data and correlate it with other municipal data to predict crime hot spots. Many police reform activists and AI ethicists have pointed out that because crime data and policing are often racially biased, so, too, are these predictive AI systems that employ this data.

Moving Forward Ethically

Companies and organizations making serious investments into artificial intelligence need to make a commensurate investment into safety, bias prevention, and related ethics. There will always be the risk of human bias carrying over into AI systems. These systems are looking for patterns—and discriminatory bias is a pattern. Similarly, there will always be safety concerns when automating systems or processes that affect people, especially in an industry like healthcare.

These risks should not deter us from moving forward with AI technology. Bias is a fact of life. Safety is always a concern. The truth is that artificial intelligence is a powerful tool for building a safer, more just world. Automating systems removes the risk of human error and improves consistency. Yes, errors and mistakes get repeated at scale—but so, too, do fixes and solutions.

Artificial intelligence can also be a tool for making better, more informed decisions about serious ethical dilemmas. Many of these issues have always existed. The decisions we are now asking AI developers to make carefully were traditionally handled in a slapdash, decentralized way out in the field. Drivers who are moments away from a collision are making split-second ethical decisions about damage mitigation. We may be less prone to second-guess the bus driver who swerves into oncoming traffic to avoid hitting a pedestrian, but the decision was made all the same. AI allows us the amazing and unprecedented opportunity to think these decisions through thoroughly in advance.

> Companies and organizations making serious investments into artificial intelligence need to make a commensurate investment into safety, bias prevention, and related ethics.

However, this does shift the dilemma from the individual to the AI developers and society as a whole. This is a profound change. People are not always comfortable having machines, or their designers, making life-or-death decisions in such a cold and methodical way, not at first. However, people can be acclimated to automation. We must show people how automation actually makes the world safer and more just. Someday passengers may be as comfortable with autonomous vehicles or automated healthcare procedures as many are with automated air travel.

Citizens are right to second-guess and question AI design choices more than the split-second decisions made by humans. Engineers and developers have the advantage of thinking these things through ahead of time. We *should* be making better choices. That is, after all, the entire point of automating these things.

The stakes are high. An individual making a bad decision has consequences for the immediate situation. An AI system making bad choices that get replicated at scale hundreds, thousands, millions, maybe billions of times is serious business. Design mistakes can have grave repercussions when we are operating at scale. We have to get these issues right.

Ultimately, we will need a global framework on AI ethics and regulatory bodies that can provide guidelines, benchmarks, and monitoring. However, exponential technologies grow exponentially. Policy does not. Policy will always lag behind these technologies. Industry cannot wait for policymakers to step up. In the meantime, industry must self-regulate. We must understand the perils and create protocols and systems for mitigating risk. We must implement robust monitoring and review processes. We must check data and insights for biases. Strong company and industry policy and thoughtful self-governance can help us mitigate these risks.

Transparency is key. AI decision-making is often conducted in a black box, especially as systems become more complicated. Machine learning can take AI systems down paths developers never imagined. We have to be ready to explain our design choices—because regulators and lawmakers might one day call on us to do so.

Decisions that these systems make will sometimes cause harm. That's unavoidable. However, AI also has the potential to save and improve lives. We are being presented with opportunities we cannot pass up. We have to move forward—yes, ethically, but also forward.

Nowhere is this truer than in the healthcare industry. We have to be ready to make the case that these technologies can save lives and improve society. That means thinking long and hard about our design choices and how to communicate them to the public. If we do the job right, we'll be able to show the world how these revolutionary technologies are making the world better for everyone. We will be able to demonstrate that there is cause for concern but not fear.

CHAPTER 6

ARTIFICIAL INTELLIGENCE AND HEALTHCARE

IN CHAPTER 4 of this book, we discussed the four basic applications of AI in industry: *automation*, *insights*, *personalization*, and *sensing*. These applications can be applied in a variety of use cases within the healthcare industry to create efficiencies, improve outcomes, and bring down costs.

Although there are countless use cases for applied AI within the healthcare industry, we will cover four broad areas:

- Automation of business processes

- Diagnostics

- Drug discovery

- Prognostics

More advanced AIs might come in the future that will have new sets of use cases. Mastery of general AI will open up whole new possibilities, such as fully automated surgery performed by robotics. This is not possible with the current generation of AI technology. However, applied AIs can now assist in surgery. No existing AI systems can perform brain surgery, but there are AI tools that can identify the best path to reach a tumor deep in the brain without damaging vital neural tissue.

Currently, AI is mostly applied in narrow ways that assist clinicians by providing insights. That may not be as futuristic as robot surgeons, but it is highly valuable to providers and the patients they serve. When these insights improve outcomes and bring down expenses, they also benefit the insurance companies trying to underwrite the cost of healthcare.

Let's look at the four primary kinds of use cases for applied AI in healthcare.

Automation of Business Processes

The American healthcare industry is incredibly complex. There are multiple sectors with numerous stakeholders. The medical sector encompasses many hospitals, nursing homes, urgent care centers, and sole or small practitioners. Clinicians, nurses, medical assistants, rehab specialists, psychiatrists, and other providers work with a variety of patients. The pharmaceutical industry has its own ecosystem of pharmaceutical companies, public and private researchers, manufacturers, distribution pipelines, and drugstore vendors. The insurance industry has a complex public and private sector working in cooperation to serve policyholders. There are just an incredible number of stakeholders involved in US healthcare.

In order for our decentralized healthcare system to function, a great deal of information must change hands between all of these actors every day. Workflow chains are long and stretch across separate institutions and back again. Think about how many different entities need access to your billing and medical records when you want to use your insurance to get a referral to a hospital specialist from your primary doctor. As medical information is particularly sensitive, records have to be handled securely as they are moved around.

All this careful transfer of information creates a huge administrative burden. Inefficiencies slow down the entire system. Workflows bottleneck easily with so many links in the chain. This can easily create cascading failures. The insurance company didn't get that medical record from the primary care physician? Then the specialist cannot perform the operation. The insurance company hasn't replied to a preauthorization request? Labs cannot be ordered. The reimbursement request didn't match the preauthorization request? The doctor doesn't get paid for the work and won't continue with treatment. A single holdup anywhere in the chain can cause things to grind to a halt. This often leaves patients or providers scrambling to figure out what went wrong.

Consider what this looks like for someone like John. When he comes down with a mysterious stomach ailment that he cannot shake for a week, John finally goes to see his doctor to get checked out. His doctor does an exam and orders several tests. When the tests come back, the doctor suspects irritable bowel syndrome. John and his doctor work to devise a treatment plan. First, the doctor needs labs before he can prescribe meds.

The doctor sends John's insurance company a preauthorization request for the labs. The doctor needs to know the labs will be covered. Each insurance company has their own set of guidelines on what services are "medically necessary" and covered. The insurance company compares the specifics of John's case against these guidelines and decides whether or not to approve the treatment. If they offer preauthorization, John can get the tests. If not, the insurance company makes an alternate suggestion.

After performing exams or procedures, the doctor now has to file for reimbursement despite already having gotten preauthorization. This has to be done at least once for every test or treatment. The insurance company then has to verify whether the claims match the original requests on the preauthorization request. Any discrepancies require the insurance company to follow up with the doctor before issuing the reimbursement. There is a whole process for addressing discrepancies and anomalies, which can also get quite complicated.

Ultimately, what may seem like a simple trip to the doctor for John sets off a long administrative process on the back end. This can require a significant amount of labor and resources for both the doctor's office and the insurance company. There are also ample opportunities for something to go wrong. If the preauthorization doesn't go through, the labs will not be covered. If the reimbursement is denied, John might get a surprise bill in the mail. Then he is the one who is left trying to figure out why his claim wasn't paid.

Not being a healthcare administrator, John isn't well equipped to determine why a preauthorization or reimbursement didn't go through. He never even knew of these processes until something misfired. John just wants his claims paid. When they aren't, he has to make the hard choice to forgo care or eat the cost himself, even though he already pays for insurance.

The inefficiency of this patchwork system is very costly to institutions. Hospitals and insurance companies bear a high administrative burden managing processes that require so much coordination and troubleshooting. This is just the price they pay for operating in a complex industry with so many actors working across institutional lines.

Ideally, we would like to eliminate inefficient processes entirely. Given the realities of the current healthcare system and the context in which healthcare organizations must operate, an immediate remedy to these woes is automation of these business processes. The flow of information can be automated easily. With the help of AI, the system can decide how to make requests and respond to queries automatically. This allows computer systems to collect the information they need to carry out business processes automatically.

Automation can be applied across all healthcare administrative institutions to seamlessly pass information from one to the next. Hospital inventories can be monitored and restocked by computer systems. Preauthorizations can be requested automatically. Customer service centers can be outfitted with first-line AI chatbots that can handle most customer inquiries. Claims can be processed entirely by computers.

Even document review can be automated. This is often an essential part of automating business processes. The healthcare industry produces an enormous number of records. Staff and providers must collect and record vitals. Doctors and nurses keep and update patient charts. Clinical notes can run dozens of pages. And that's just direct care. The rest of the industry also produces records. Insurance companies and hospitals produce forms, guidelines, and written policy. Research institutions and pharmaceutical companies conduct studies and put out data. Every last organization produces a trail of administrative forms and records for every single business process.

Whole armies of administrators are paid to sift through these records. Records get pulled and checked and referenced. Every little decision that is made relies on records, often many of them. Medical coders are employed by doctors to make sure that providers and institutions are reimbursed for services rendered. This is not simple data entry. Coders are involved in document review and synthesis of information. They are making decisions about what data to input.

Historically, that decision-making process made the coding process harder to automate. Someone had to review unstructured data, consider context, and make a series of complex decisions on how to interpret and code the case. Now that applied AI has improved, this process can be largely automated. The current generation of AI can—within certain constraints—read many documents much in the way that humans would. Natural language processing allows AIs to review records, doctors' notes, patient history, and other documents. This lets AIs take in unstructured data and work with it. AI systems can now do much of a medical coder's job automatically.

The key here is that AI allows computers to engage in simple decision-making to offload simple repetitive tasks to computers. As we saw in previous chapters, these systems are just as accurate as humans. Automated claims processing is now about 95 percent accurate, the same as humans perform, only faster. This should come as no surprise. At Anthem, we train AIs on how our actual human processors would handle cases. Edge cases get flagged for manual review. Because most claims are straightforward, the AI system frees up human resources to focus on cases that require special consideration. The result is an administration system that is much faster, more accurate, and far less costly.

The healthcare industry has already automated many processes. However, we need to go all in on automation. At Anthem, as an AI-first company, we automate everything we can and are constantly looking for ways to improve systems. AI technology is advancing rapidly. Processes that cannot be automated today might be automatable tomorrow. Keeping up with the latest exponential technology is a major endeavor. But the returns for doing so keep getting better as the technology grows exponentially.

Diagnostics

Artificial intelligence can make decisions on more than just administrative processes. Applied AIs now actually assist with clinical decision-making and diagnosis. As you may recall from chapter 2, IBM's Watson Health AI system struggled to meet expectations in this regard at the University of Texas MD Anderson Cancer Center. The failure of that project should not be seen as an indictment of the diagnostic power of AI Watson Health. The concept was simply applied too early and too broadly.

AI systems absolutely can assist with diagnosis and decisions about treatment plans. The operative word here is *assist*. AI must be applied in an appropriate manner given the constraints of the current technology. The technology is not ready to take over for clinicians. However, AI tools can make clinicians better at their jobs.

These tools are able to rapidly collect and synthesize relevant data and make clinical recommendations based on full available information. Doctors typically act on intuition when coming up with treatment plans. Over years of practice, they develop a feel for what works and what doesn't. They don't have the capability to access, assess, and synthesize *all* of the relevant data on a case, including years of patient outcomes. AIs absolutely can do just that. They are able to consider a patient's entire medical history and compare it to similar cases. They can factor in all relevant studies ever done. They can consider the likelihood that a therapy will work for a particular patient with a methodological precision and thoroughness that humans cannot. This level of granular information about what has worked on whom allows doctors to find the best treatment pathway for a particular patient. It allows for personalized patient care planning.

The difference between a human diagnostician and this kind of AI system is the difference between a seasoned master chef cooking by feel, pinching salt and eyeballing ingredient proportions, or an industrialized cooking apparatus producing the same optimized outcome every single time.

A better analogy might be that current AI systems are more like the latest high-tech cooking apparatus in the hands of a master chef. Clinicians use applied AI systems as diagnostic tools. The AI can return insights and recommendations, but clinicians still review the results and draw their own conclusions. They consider context and display wisdom that applied AI systems do not. Someday we may have powerful general AI that can perform as autonomous diagnosticians. For now, AI has a supporting role assisting clinicians.

However, the value of that supporting role cannot be overstated. AI allows clinicians to perform more effectively and efficiently. Applied AI is making good doctors great. Clinicians have never had such powerful diagnostic tools.

These tools don't just return diagnoses. They also assist in developing personalized treatment plans. Consider the potential of AI in devising epilepsy treatment plans. Epilepsy is a complex disease for which there's no known cure. However, we do have several medications that can control the disease. These medications have different efficacy and tolerability in different patients. Clinicians must help patients find the medication that best controls the condition most tolerably.

This is typically done through trial and error. The doctor prescribes different combinations of medications at various dosages until the patient's epilepsy is controlled without inducing intolerable side effects. The trial and discovery process can take weeks, months, or even years. During this time, the patient is at risk of experiencing seizures and negative reactions to the medications. There is also no way to know when a particular therapy is good enough. The patient has no way to know if another therapy might work better unless they have tried them all.

Artificial intelligence can streamline this process by personalizing the treatment plan to the patient. Doctors can use AI to match drugs to patients based on their individual profile. Genetic, microbiome, health history, and other personal factors can be compared to those of other patients to find the therapies that are most likely to be effective and tolerable. The therapy most likely to work is tried first. Once an effective and tolerable therapy is found, the odds that another therapy will be considerably better are substantially lower than if you had just been trying them at random.

This level of personalization is not something doctors can accomplish without AI. Epilepsy is a complex condition. There are many variables involved. It takes an AI system capable of correlating a patient's individual data with a vast trove of historical outcomes for patients on different therapy plans. Synthesizing all that data into a recommended treatment plan personalized to an individual patient is not something humans can do without AI assistance.

This is not some hypothetical technology. Companies and organizations are partnering to build these kinds of AI platforms now. Anthem is working with doc.ai, an artificial intelligence company working in medical research, to create an epilepsy database that the AI tool can use to make personalized treatment plans. Their app allows for the tracking of patients' symptoms and responses to various drug therapies. The system will ultimately track seizures, symptoms, drugs, dosages, and side effects as well as other patient data related to genetics, biometrics, family history, lifestyle choices, and more. This data allows the AI system to look for correlations and insights.

Our project deals with epilepsy, but the app has many other ongoing trials for different conditions and diseases. As these systems are refined and see greater adoption, they provide better and better insights. Someday soon, they will be central to how doctors develop personalized treatment plans for all kinds of conditions.

Drug Discovery

Pharmaceutical companies can also use applied AI to discover new drugs and other therapies. The drug development cycle is long. Pharmaceutical or life sciences companies start by researching a particular condition. Then they try to identify potential therapies and attempt to guess which one will be the best. When a suitable candidate is found, the new drug has to go through multiple rounds of animal testing followed by multiple rounds of human trials. The average drug now takes twelve years and more than a billion dollars to bring to market—and that's just from the time of the FDA application through Phase 4 trials.[16] Some drugs take longer and cost several times as much.

This difficult process is meant to protect the public from dangerous drugs, which it does well. The downside is that it is slow, expensive, and highly inefficient.

We need safe drugs—but we need them at better prices.

Here is where AI comes in.

---

16   Thomas Sullivan, "A Tough Road: Cost to Develop One New Drug Is $2.6 Billion; Approval Rate for Drugs Entering Clinical Development Is Less Than 12%," *Policy & Medicine*, last updated March 21, 2019, https://www.policymed.com/2014/12/a-tough-road-cost-to-develop-one-new-drug-is-26-billion-approval-rate-for-drugs-entering-clinical-de.html.

Getting a drug to market is the only way to see a return on the massive investment of clinical trials. Currently, many therapies fail in the discovery process. Fewer still actually make it to FDA approval. Drugs abandoned during clinical trials can cost pharmaceutical companies millions or billions of dollars with zero return on the investment. Cutting down on the number of failed attempts has the potential to massively decrease development costs.

AI performs efficiently here. Scientists can use AI to simulate drug trials. Most new drugs belong to a family of known drugs. We understand how these molecules work and how different substitutions and alterations are likely to affect the human body. This allows researchers to model the efficacy and side effects of a potential new drug before ever manufacturing it in a lab. Drug companies run simulated trials first and take only the best performers into clinical trials. This allows the pharmaceutical industry to pursue the best drugs—and bring only the best drugs—in actual trial. Researchers tweak the drug in the computer models as much as they can first before the drug ever sees a physical lab.

When the drug is ready for real-world clinical trials, AI technology can reduce trial costs by removing the need for a control group. Outcomes in the experimental groups can be compared to data from a database of medical records. This would save researchers the expense of running the experiment on the controls. They would need only to perform the trials on the experimental group.

This isn't just cheaper—it's also a better way of putting together a control. These virtual control groups are *not* simulations. They utilize real records from real patients. Assembling a control group in this way produces a data set that is no different from what you get from a traditional control group.

Control groups could be as broad as the entire general population. They could also be tailored to match the needs of the study. Experiments could compare results to a defined subset of the population that matches the experimental group.

We already have the data necessary to create virtual control groups. We just need to capture and make it available as needed. This is already starting to happen. The US Food and Drug Administration established the Sentinel System to leverage preexisting health data to monitor the safety of medical products. Part of that effort involves *synthetic control arms*, as they call them, which allow trial results to be compared against real patient data. When deployed at scale and widely used, this will hasten drug discovery while reducing costs.

Prognostics

Artificial intelligence doesn't just allow for better diagnostics. The insights gleaned from data also make it possible for us to make predictions about the future. When applied to health, this allows for a clearer understanding of disease emergence and progression, which can be used to provide better preventative care.

This can be done at both the population and the individual levels.

POPULATION LEVEL

At the population level, AI tools can predict the outbreak and spread of new diseases. This was how the disease caused by the novel coronavirus, which came to be known as COVID-19, was first detected. The CDC issued an alert about a flu-like respiratory illness circulating in Wuhan, China, on January 6, 2020. The World Health Organization issued its own alert three days later. However, these were *not* the first warnings of the outbreak. BlueDot, a Canadian health monitoring platform that uses AI algorithms to scan for new outbreaks, detected an unusual cluster of pneumonia cases possibly related to a Wuhan wet market a full week before the CDC or WHO issued their alerts. Other privately held AI systems also picked up on the outbreak more quickly. The Boston Children's Hospital ran an AI system that detected the outbreak before the major regulatory bodies. So did Metabiota, a private company based in San Francisco that monitors infectious diseases and outbreaks.

AI systems were also able to predict where COVID-19 would spread next. BlueDot correctly predicted that new hot spots would emerge in Bangkok, Seoul, Taipei, and Tokyo within days based on analysis of travel data. Throughout the pandemic, epidemiologists continued to employ AI to track the virus as it spread around the globe. This modeling was crucial to understanding the global pandemic in the absence of adequate testing. Being able to test everyone directly to know where the virus was spreading would have been ideal. Unfortunately, this wasn't possible given the novel nature of the new coronavirus and the speed at which it spread. The testing capacity was simply not in place fast enough. This clouded our understanding of the virus and its progression.

Epidemiologists got around the problem by employing AI to model spread, transmissibility, and fatality rates. Collecting data on outbreaks and outcomes allowed them to draw important insights about the virus. Epidemiologists were able to accurately forecast death rates, study best practices, and predict future hot spots where the virus would erupt. The analysis revealed the potential seriousness of the virus in the absence of tests. These insights were crucial to decision-making on tough policy choices, such as whether to implement citywide lockdowns.

INDIVIDUAL LEVEL

As the pandemic progressed, we learned more about who was most at risk and why. Epidemiologists discovered that age and certain comorbidities put people at higher risk of serious infection, hospitalization, and death. Sometimes these insights were unintuitive. Thinking we were dealing with a standard respiratory infection, most doctors assumed that asthma, COPD, and other lung issues would put patients at greatest risk. What they discovered was that age and cardiovascular problems were the larger risk factors, pointing toward a more complicated disease progression. This explained why many patients were dying of heart attacks or strokes.

As the pandemic progressed, doctors were able to use these insights to develop an intersectional understanding of risk factors. This informed our understanding of which groups needed to implement the strictest preventative steps due to their specific risk. It also led to more effective treatment guidelines for the infected. The population-level data provided insights into individual risk and allowed people to receive more personalized care. Patients facing respiratory distress and developing acute respiratory distress syndrome might be put on oxygen or a ventilator while those exhibiting cardiovascular issues might be screened and treated for the development of blood clots. The treatment was tailored to the individual.

These same strategies can be applied to chronic diseases, such as diabetes. IBM researchers are currently working on an AI-driven screening tool to detect early cases of type 1 diabetes. Although the United States sees forty thousand new diagnoses each year, there is no screening tool. Many early cases go undetected until someone ends up in the emergency room. IBM partnered with the Juvenile Diabetes Research Foundation, a nonprofit research institution, to map the presence of type 1 diabetes antibodies in thousands of test subjects. Their AI tools recognized patterns in the data that allowed for predictions about the development and progression of the disease. This allowed for better screening tools and protocols. The population-level findings provided insights into individuals based on their specific risks and markers. This gives us a better idea of whom to test and how often to do so. Patients can be informed of their risk and, if they already have the disease, how far along they are in its progression.

The same thing can also be done for other chronic diseases. The Computer Science and Artificial Intelligence Laboratory at MIT recently developed an AI tool that can predict the development of breast cancer up to five years in advance.[17] Mammograms are scanned for precancerous formations. These visualizations are compared to those of other patients with known outcomes. This provides insights that can help predict the likelihood that someone will develop cancer. The tool already outperforms current screening models, which have often been criticized as ineffective. Future improvements on the algorithm might include assessment of genetic risk. Historically, that risk was assessed by family history. By observing genetic markers and mammograms directly, AI tools avoid the bias implicit in preconceived notions of risk. The machine is able to assess risk better by observing objective markers in cases with known outcomes. This is a sounder methodology than making guesses about someone's "family history," which might include only a few closely related women in the sample.

Similar research is being done on Alzheimer's. A resident at the Department of Radiology and Biomedical Imaging at UC San Francisco used machine learning to analyze PET scans of the brains of

---

17 Darrell Etherington, "MIT AI Tool Can Predict Breast Cancer up to 5 Years Early, Works Equally Well for White and Black Patients," TechCrunch, June 26, 2019, https://techcrunch.com/2019/06/26/mit-ai-tool-can-predict-breast-cancer-up-to-5-years-early-works-equally-well-for-white-and-black-patients.

Alzheimer's patients. By measuring subtle differences in brain glucose levels and other biomarkers, the AI system was able to predict which other patients would eventually develop Alzheimer's—up to six years in advance with an accuracy exceeding 90 percent.[18]

Researchers at the University of Oxford are using machine learning to identify biomarkers that can predict risk for heart attacks and arteriolosclerosis.[19] They believe that they can predict heart disease five years in advance of development and typical diagnosis. Cardiovascular health, like many chronic illnesses, responds well to early intervention. Arteriosclerosis is hard to reverse but easy to avoid. Preventing arteries from hardening in the first place is the best possible outcome.

It's also the most affordable treatment. This is where prognostic AI really has the potential to reduce costs. As we know, preventing chronic illnesses is an order of magnitude less expensive than treating them. But we have to get people into preventative treatment before they develop the disease. Being able to tell patients with a high degree of certainty that they *will* get cardiovascular disease, not just that they *could*, is a strong motivator to make lifestyle changes.

Personalized prognostics make future health outcomes more real to patients. The surer the prediction, the more motivating. Remember how the lifetime odds of dying in a car accident for Americans is about one in one hundred? This is a disconcerting fact, but it is unlikely to dissuade most people from using a car. Now tell someone that their odds of dying in a car accident next Tuesday are 50 percent, and they will very likely stay home on Tuesday!

---

18   Dana Smith, "Artificial Intelligence Can Detect Alzheimer's Disease in Brain Scan Six Years before a Diagnosis," University of California San Francisco, January 2, 2019, https://www.ucsf.edu/news/2019/01/412946/artificial-intelligence-can-detect-alzheimers-disease-brain-scans-six-years.

19   Andrea Downey, "Artificial Intelligence 'Predicts Fatal Heart Attacks up to 5 Years in Advance,'" DigitalHealth, September 3, 2019, https://www.digitalhealth.net/2019/09/artificial-intelligence-predicts-heart-attacks.

While we obviously cannot provide that kind of certainty narrowed to a single day, AI insights are getting better and better at predicting outcomes for specific individuals. Vague statements about an elevated risk of diabetes or heart disease might not persuade someone to start eating better and exercising more, but what about a 95 percent risk within ten years? This is more likely to get someone to adjust their behavior and make lifestyle changes. This requires sacrifices in the present for the benefit of a distant future, but the less hypothetical negative future outcomes become, the easier such sacrifices are to make.

Prognostic AI tools become more accurate as we track more variables among the population. The more granular our health data, the better we can quantify specific risks in individuals. Consider cancer. Improved genome sequencing can reveal a person's inherited cancer risk. Looking at comorbidities that increase cancer risk, such as certain viral infections like HPV or hepatitis C, gives us a clearer picture of risk. We can also calculate the likelihood that a patient will contract such a viral disease based on lifestyle choices. Injection drug use and unprotected sex raise the risk for these infections, which raise the risk for cancer. We could then also look at environmental factors, such as economic income and place of residence, that influence the risk of behaviors (e.g., addiction) that lead to these infections, which would therefore also correlate with the risk of certain cancers.

As you can see, we can go very far down the rabbit hole in search of more detailed and granular risk factors. There are probably unlimited variables related to cancer risk that we could potentially track. AIs are even likely to discover new risks as they analyze data and return new correlations. These tools employ an adaptive model. They use machine learning to improve their own algorithms. Our AI algorithms will become immensely powerful when deployed at scale so that they can gather patient data on the entire population. This gives us more data to compare an individual's risk against. Our ability to make more certain predictions improves with more data and insights.

Creating a clearer picture of risk will help patients and providers take corrective action *before* conditions develop. This is being done now, but it hasn't been adopted at scale across the whole healthcare system. It needs to be. Doing so will lower costs and improve quality of life. Artificial intelligence can provide personalized insights into lifestyle choices and other factors that lower an individual's specific risks.

Currently, most health guidelines are based on insights about the general population. These kinds of insights tend to confirm what we already know. Eating healthily, exercising, wearing sunscreen, getting screenings—these things all improve general health. The truly revolutionary power of AI will be in finding patterns that tell us about an individual's specific risks. This will allow for personalized preventative care.

> **Better diagnostics and therapies are great, but the biggest room for improvement in health outcomes and cost reduction is in prevention. AI-driven prognostics make that possible.**

When individuals know their actual risk, they can take targeted mitigatory steps. Specific lifestyle risks can be altered. Genetic risks can sometimes be compensated for through increased testing, behavioral modification, or prophylactic medication. Powerful prognostic capabilities that can give someone their true risk with a high degree of certainty would provide more incentive to take targeted proactive steps to stay healthy.

A clearer picture of an individual's risk and costs doesn't just incentivize patients to seek preventative care. More predictable lifetime costs for each patient also give insurance companies the incentive to push for better preventative care. They will know exactly what a high-risk patient is going to cost them. This will help insurance companies enact targeted mitigatory measures. High-risk policyholders could be better rewarded for making smarter choices about their health. They could also be "taxed" on premiums for failing to do so. This is a carrot-and-stick approach that would align the interests of policyholders with their insurance companies.

Aligning incentives in this way is crucial. Allowing chronic conditions to develop and advance puts a greater strain on the healthcare system than anything else. This is becoming increasingly true as we develop advanced therapies that, while effective, are often quite expensive. Better diagnostics and therapies are great, but the biggest room for improvement in health outcomes and cost reduction is in prevention. AI-driven prognostics make that possible.

The benefit to patients will be substantial. Consider someone like our hypothetical John. His failure to control his metabolic syndrome resulted in full-blown diabetes. This was largely the result of John rationalizing his behaviors and downplaying the risk. But what if he had known with greater than 95 percent certainty that he would develop diabetes in only ten years? This knowledge very well could have been enough to alter his behavior.

Preventing that diagnosis would have made a major impact on his quality of life. In economic terms, the diagnosis was devastating. Eating a healthy diet and working out would have cost him next to nothing. Now that he is forty-five and diabetic, John is looking at paying much more for a lifetime of reactive care. The American Diabetes Association puts the average annual cost of diabetic care at

$9,601 and growing. There is also the compounding cost of cascading health problems resulting from diabetes. The average diabetic spends 2.3 times more on healthcare than those without a diabetes diagnosis. In 2017, the average diabetic was spending $16,752.[20] That figure is only growing.

Given these costs, John's diabetes might easily cost him in excess of a million dollars over the course of a lifetime, and that's before adjusting for future inflation. He won't just be paying for insulin, syringes, and pumps. There is also the cost of intermittent doctors' visits and hospitalizations. His vision will cost more to protect. Everything related to healthcare will cost him more.

And that's just the monetary cost—the greatest cost of all will be his lower quality of life. You cannot put a price on your health, not really. John will face a series of related health consequences that will leave him much sicker, exhausted, and more miserable than he would otherwise be without diabetes.

All of that could have been avoided if someone would have looked John in the eye and said, authoritatively, "John, you *will* get sick. This is what will happen within five years. This is what it will cost you. And here is what you can do to prevent it."

Or his insurance company could have given him the ultimatum: "This is what it will cost us to treat you, one million dollars, and unless you make these changes now, we are going to start charging you for that now in installments." This might have been all John needed to get his act together. He would have only needed to make changes to his diet and activity level while keeping up with monitoring. Maintaining that effort would still have taken work, but, in so doing, John would ultimately be healthier. He would have avoided the complications and expense of a chronic disease that he *knew* was coming.

---

20   American Diabetes Association, "Economic Costs of Diabetes in the U.S. in 2017."

And John isn't the only one to benefit. His insurance company would benefit by having lower and more predictable costs to underwrite. Society would benefit from the reduction in medical expenses, which means lower premiums and total costs for everyone. Reducing the burden of chronic illness by shifting to preventative treatment benefits us all.

Effectively, what AI-driven prognostics do is help align the patients' incentives with their future interests. By making the future clearer, patients like John are incentivized to make better, more forward-looking decisions. They still have to put in the effort to make the necessary changes, but they can no longer easily hide behind probabilistic uncertainty. It's no longer so easy to rationalize poor choices.

Prognostics also help align the incentives of insurance companies with the interest of policyholders. The shift to preventative care benefits those underwriting the cost. This makes it easier for the industry to justify covering preventative care. This also realigns the interest of policyholders with the collective, who will all enjoy lower premiums when more people make better lifestyle choices.

However, some issues would still remain. Individual insurance companies still have to contend with the problem of policyholders switching insurers. This still disincentivizes investments in preventative care that some other insurance company will reap the rewards of.

However, there is another novel exponential technology that can address this issue, as well as others, such as the need to securely transfer sensitive information through the medical system. That technology is blockchain.

Let's now turn our attention there.

CHAPTER 7

BLOCKCHAIN EXPLAINED

MOST PEOPLE KNOW BLOCKCHAIN as the technology behind Bitcoin. In fact, blockchain technology is so associated with the cryptocurrency space that the two are often mistaken as synonymous. They are not. Blockchain is its own exponential technology. Cryptocurrencies are merely one application of blockchain. There are many others. Blockchain has numerous business applications that could help revolutionize the healthcare industry, which we will explore in the next chapter.

First, we need to understand what blockchain is and is not.

At its core, blockchain is simply cryptographically secured ledgers that are distributed across a peer-to-peer network. Blockchains are a kind of distributed database. This is not to be confused with cloud databases, which are centralized on servers and accessible to users online. Blockchains do not exist in any one place. There are no central servers. The data is distributed across a network of users serving as nodes for authorizing other users, sharing data, and validating changes to the blockchain.

Blockchains are referred to as ledgers because they contain an *immutable* record of all transactions and changes. Blockchains are updated with the sequential addition of new "blocks" of data that contain changes and updates. Authorized users can go back and observe the entire history of the blockchain. These blocks are approved by the network through some form of consensus mechanism. Cryptography secures the blockchain by ensuring that only authorized users can interact with the blockchain and that users are who they say they are. In this way, blockchains provide an immutable shared record of truth about a history of peer-to-peer transactions.

The Rise of Cryptocurrency

The earliest work on blockchain-like technologies was rooted in early e-commerce. In the late 1990s, online transactions between buyers and sellers were mediated and settled by a third party, just as they are now. However, payment authorization was painfully slow at first. Banks and credit card companies took several minutes—sometimes even hours—to validate online transactions. This lag was obviously a problem.

Technologists began imagining ways that buyers and sellers could remove the intermediary and transact directly. No intermediary, no waiting on them to validate transactions. The internet was revolutionizing the direct exchange of information—could financial transactions be handled in a similar peer-to-peer manner?

Handling monetary transactions presents unique problems that the exchange of information often does not. Invalid information can be reviewed and discarded. Monetary transactions have to be verified as valid when processed. Otherwise, scammers could abscond with the funds associated with an invalid transaction. Online retailers needed a mechanism for ensuring that electronic monetary transactions could be trusted in order to make direct transactions with buyers. In the absence of this mechanism, an intermediary or escrow service is necessary, complete with all the aforementioned problems.

E-commerce companies mostly abandoned development in this area once banks and credit card companies improved transaction times. Today, there are several intermediaries offering virtually instantaneous online transactions, from banks to credit card companies to private payment services like PayPal and Venmo. E-commerce companies stopped trying to make direct transactions with their customers and went back to working on other aspects of their businesses. However, technology enthusiasts continued to look for ways to conduct direct online transactions without intermediaries.

In 1997, technologist Nick Szabo published "The God Protocols," an article that imagined how transactions without an intermediary should look.[21] The optimal protocol would function like a benevolent deity that collected information from each party and settled the transaction fairly. This would allow for trustless exchange of currency, information, or anything else. A so-called God protocol would accept a debit from a seller and credit from the buyer, ensure that they were accurate, and settle the transaction without an intermediary. The protocol would have to be omniscient and godlike in order to remove the need for an intermediary.

The following year, Szabo developed the concept of bit gold, a hypothetical digital currency whose value and security would not depend upon trust in a third party. Instead, a peer-to-peer network would maintain a ledger of debits and credits. Although the idea was never implemented, due to various unsolved problems, the proposed

---

21   Nick Szabo, "The God Protocols," Satoshi Nakamoto Institute, 1997, https://nakamotoinstitute.org/the-god-protocols.

architecture was a precursor to Bitcoin. (In fact, many have speculated that Szabo is the true identity of Satoshi Nakamoto, the pseudonym of the anonymous creator of Bitcoin, although Satoshi's true identity has never been substantiated.)

Szabo and others worked on various digital currency projects over the years, but as with all nascent exponential technologies, progress was slow at first. A decade passed before the anonymously published whitepaper "Bitcoin: A Peer-to-Peer Electronic Cash System" appeared in 2008.[22] The open-source software soon went live, and in January 2009, the very first Bitcoin was "mined." This was the first true working blockchain.

And work it did. Over the next decade, Bitcoin saw an exponential increase in adoption, value, and physical infrastructure supporting the network. The cryptocurrency ecosystem has grown into a worldwide phenomenon with the expansion of large crypto exchanges. Established companies have already moved into the space. Facebook even announced in June 2019 that it is developing its own cryptocurrency, Libra. Ripple Labs, creators of the XRP cryptocurrency, are developing various distributed ledger technologies for the financial industries. The company now works with over one hundred banks.

Bitcoin was a direct answer to the 2008 financial crisis. Bitcoin true believers feel that cryptocurrency offers an alternative to the fractional reserve banking systems that had nearly failed along with the global financial sector. Bitcoin provides a way to settle transactions between peers without any intermediary. At scale, which would not be possible without the exponential growth we are seeing in the blockchain space, peer-to-peer transactions remove the need for centralized regulators or intermediaries. Bitcoin (and other cryptocur-

---

22   Satoshi Nakamoto, "Bitcoin: A Peer-to-Peer Electronic Cash System," 2008, https://bitcoin.org/bitcoin.pdf.

rencies) at least threaten the entire banking and financial sector with obsolescence.

Here's how it works: The Bitcoin peer-to-peer network is based around a distributed ledger, the Bitcoin blockchain, that logs Bitcoin addresses and how much Bitcoin they hold. Each address is associated with an algorithmically linked pair of keys. The *public* key is basically the address itself. The entire network can see the public key and any associated Bitcoin. The *private* key proves ownership over the public key. The blockchain protocol can deduce the public key from the private key. However, the private key cannot be deduced from the public key. This is done algorithmically so that the private key never has to be shared with others, which is part of what makes blockchain so secure.

Bitcoin is created by mining for it. Network users compete to be the first to solve complex mathematical problems that allow them to create and verify the transactions on the next block, which is then added to the blockchain and verified by the peer-to-peer network. For their troubles, the user who solves the problem first receives a reward of Bitcoin. The mathematical problems scale up in difficulty proportionally with the computational capacity working to solve them. In the beginning, Bitcoin could be mined on a personal computer. Today, as Bitcoin has appreciated in value, vast mining farms work around the clock to solve these problems.

This system, known as proof of work (PoW), is the consensus mechanism for adding blocks to the Bitcoin blockchain. While the mechanism has downsides—namely, the amount of electricity spent on computation that scales up with the price—PoW is a proven consensus mechanism that allows for trustless transaction. Requiring computational work ensures that no one can easily take over the network. The vast amount of computing power and electricity makes unilaterally taking over the network and forcing consensus practically impossible.

Users need not mine Bitcoin to make transactions on the ledger. They can simply have Bitcoin transferred to (and from) their address for the cost of a small fee paid to the miner that verifies the block on which the transaction is contained.

It is important to understand that the miners are *not* a centralized authority. Miners are needed to validate transactions, but anyone can lend computing power to mining. Commercial mining farms carry out much of the work of validating consensus, but open competition for mining rewards keeps the network distributed and secure. This works so well that the Bitcoin blockchain has never once been hacked or compromised. Hacks can be socially engineered, and many people and exchanges have been breached, but the Bitcoin blockchain itself has proven entirely secure so far.

Blockchain Basics

Not all blockchains are structured exactly like Bitcoin. There are different types of blockchains with different structures. However, all blockchains do share three defining characteristics:

1. Distributed peer-to-peer networks

2. Cryptographically secured

3. Immutable and only updated by a consensus mechanism

This is what separates blockchain from other more traditional data architectures, such as standard centralized databases.

DISTRIBUTED PEER-TO-PEER NETWORKS

Decentralization is the whole point of blockchain. Blockchains make it possible for users to make direct peer-to-peer transactions without any central authority or intermediary. The ledger is distributed across the network. Users act as nodes for storing data on the blockchain. Users verify each other's transactions. The whole network is peer-to-peer.

Blockchains are not limited to logging financial transactions. Blocks can contain any data, including images, videos, audio files, and even working code. Anything that can be stored in a traditional database can also be logged on a blockchain and accessed by authorized network users. This allows blockchains to serve as distributed databases. Data and records that exist in many different information systems spread across the globe can all be shared and interacted with on the blockchain.

This is inherently more secure than a centralized database, which can be hacked in one go. Blockchains don't exist in any centralized place that can be compromised. "Hacking" the entire blockchain would require knowing each and every private key associated with every bit of data. Furthermore, in addition to security, blockchain improves both accessibility and continuity of records because all updates can be seen forever on the chain.

> Blockchain improves both accessibility and continuity of records because all updates can be seen forever on the chain.

CRYPTOGRAPHICALLY SECURED

None of this would be possible without cryptography. This is not about encryption. The data on a blockchain can be encrypted for privacy and security. However, it does not have to be. Many public blockchains, including Bitcoin, are not encrypted at all. The entire history of the blockchain can be seen by anyone.

What makes blockchains cryptographically unique are the algorithmically paired keys. The keys are technically the *opposite* of encryption. Encryption locks data so that it cannot be deciphered. In blockchain, the private key "unlocks" the public key, thus proving identity or authorized access. By proving ownership of the public key, the private key can function as an unforgeable signature.

All of this is done without the need to actually exchange keys or any other passwords, as would be necessary with a traditional information system, which makes blockchain inherently more secure. Of course, nothing is 100 percent secure. Private keys are still vulnerable to social engineering and phishing. They can be stolen if written down somewhere or saved unencrypted. Computer systems feeding into the blockchain can still be compromised. But the blockchain architecture itself is inherently secure because the entire architecture is designed around the control of identity and access.

These cryptographic techniques allow blockchain participants to make transactions with each other directly without first having to establish trust between parties.

CONSENSUS MECHANISMS

Blockchains serve as a shared *history* of truth because they are immutable. Once new blocks are added to the chain and confirmed, they can no longer be altered or deleted. Updates must be made on the next block. The entire history of the ledger can be traced back down the blockchain forever.

In order for the blockchain to serve as a *trusted* record of truth, the network must be able to trust that new blocks are valid. This requires some consensus mechanism for validating new blocks. The original consensus mechanism for public blockchains, as seen with Bitcoin, is the PoW algorithm, in which miners compete against one another to solve cryptographic puzzles. The first miner to solve the puzzle gets to confirm the next block of transactions. The network verifies that the solution is correct, and the block is then added to the chain. (The mathematical puzzles are intentionally difficult to solve by brute force, but solutions are easily verified, which makes them ideal for cryptography.) This system works because it requires computational effort (and associated electricity costs), then generally rewards the miners for their effort. The work gives network participants a stake in the blockchain and thus secures the chain against denial-of-service attacks or easy takeover.

PoW is only one consensus mechanism that happens to work well on permission-less public blockchains without encryption. While very secure, the trade-off is that the network costs scale with usage. However, not all blockchains are public, permission-less, and unencrypted. In many business applications, permissioned blockchains that restrict access and employ encryption are more appropriate and easier to maintain. Alternative consensus mechanisms exist, such as proof of stake (PoS), which reaches consensus not through proof of work but through proof of stake in the system. Blocks can be validated by consensus from enough peers who have sufficient buy-in, standing, time on the system, or meet some other predefined criteria. Such a consensus mechanism doesn't require much computational work at all.

The consensus mechanism should match the purpose of the blockchain. Proprietary permissioned blockchains probably need only a PoS algorithm. Permission-less blockchains are often appropriate for tying these systems together and making them all interoperable on one "master" blockchain. As we will see in the next chapter, such a layered system can be used to bring decentralized networks together efficiently while allowing for the exchange of information without first needing to establish trust between all parties.

Beyond Cryptocurrency—What Else Can Blockchain Do?

Broadly speaking, blockchain is useful for two things: *authentication of information* and *value transfer*. Cryptocurrency blockchains take advantage of both abilities. The Bitcoin blockchain authenticates users and transactions so that value can be transferred securely in the absence of an intermediary or any preexisting trust.

Financial transactions are not the only data that can be authenticated by blockchain. All kinds of data can be logged on blockchain and authenticated. For example, records put on the blockchain and tied to a user's identity would allow for secure authentication of information or identity. Files can be forged. Blockchain private keys cannot. This allows for the authentication of records and the management of identity.

Just because any data can be put on blockchain doesn't mean that it should be. Blockchain has gotten a lot of buzz recently. The buzz sometimes crosses into hype. Blockchain is not a straight improvement on traditional databases. Blockchain has advantages and disadvantages. The enhanced security and privacy of blockchain comes at the cost of efficiency. This is especially true of large public blockchains utilizing a PoW consensus mechanism.

Not all problems require a blockchain solution. The hit to efficiency is only worthwhile in certain contexts. Centralized databases and spreadsheets are simple but often quite effective data architectures.

They are frequently the optimal data architecture. Blockchains are only worthwhile when their unique features offer an advantage.

Whether or not to implement a blockchain—and what kind—must be decided on a case-by-case basis. Generally speaking, blockchain architecture is most appropriate for building a shared repository for logging transactions between multiple parties who need to be able to trust the validity of transactions and the security of the system in the absence of intermediaries. Databases often can serve the same purposes. However, blockchain makes more sense when decentralization is required. Blockchain can verify and authenticate data coming from multiple parties without having to employ intermediaries to manage or facilitate transactions. This is especially true when the information being exchanged is of a sensitive nature and security and trust are more important than system efficiency.

This is not to say that blockchains are inefficient. Blockchains are structurally more complex than databases and often computationally demanding. However, blockchain protocols frequently remove the need for many human facilitators in the exchange of information. They provide a framework for automating transactions that require a high degree of trust and security. In some cases, the cost of running a blockchain may well be lower than the administration costs of managing a massive database.

With these advantages and disadvantages in mind, let's now consider several blockchain use cases in the public and private spheres.

BLOCKCHAIN-ENABLED LOYALTY PROGRAMS AND COMPANY CURRENCIES

Cryptocurrencies don't have to necessarily be for general use. Companies developing proprietary cryptocurrencies (such as Facebook is doing with Libra) might one day allow for the realization of the e-commerce dream of the nineties: seamless direct transactions between buyers and sellers. This could be utterly revolutionary if company cryptocurrencies were made interoperable. The monetary system could be decentralized. The economy is already decentralized—why shouldn't the monetary system that supports it be as well?

Critics might think this is utopian—or, depending on your political views, perhaps dystopian—and dismiss the notion that a network of decentralized private currencies could ever eclipse government-issued currencies. Of course, many people assume that the dollar is still backed by physical reserves. This hasn't been true since President Nixon decoupled the dollar from physical gold in 1971. The fiat currencies issued today are only valuable because we have faith in the governments issuing them. The US dollar is basically an instrument of trust. There is no reason that people wouldn't place the same faith in companies that have built a strong reputation. In fact, there's more reason to believe they would, as companies actually have competitors. Federal governments do not.

After all, this is why people buy stocks. Company-issued cryptocurrencies might blur the distinction between currency and assets, which would give more everyday people exposure to wealth-building assets. Companies that perform would see their cryptocurrencies increase in value. These currencies could even be traded, much like stocks. This would require a robust regulatory framework, of course. We see the beginning of this now with the SEC ruling that initial coin offerings (ICOs) should be treated like securities. This slowed the development of the ICO ecosystem, which was probably a good thing, as many of the companies issuing them existed mostly on paper. However, as mature companies start issuing their own cryptocurrencies, we are likely to eventually see a more decentralized and democratized monetary system. The politicization of central banking policies and government manipulation of the money supply may one day be a thing of the past.

Such a future will take time to develop and mature. In the meantime, companies can use blockchain to administer sophisticated loyalty points programs, with points that are persistent, tradeable, and interoperable. Such programs might slowly, with time, evolve into a more distributed and decentralized monetary system.

SMART CONTRACTS

The second most popular cryptocurrency is Ethereum, which runs on a decentralized open-source blockchain with advanced smart contract functionality. Smart contracts are simply transaction protocols that execute automatically when the terms of an agreement have been met. Vending machines are sometimes considered the original smart contract. Put in your nickel and the machine will give you a Coca-Cola automatically. The creators of Ethereum actually saw Bitcoin as a rudimentary form of blockchain-enabled smart contract. Bitcoin are only transferred from one address to the next when certain conditions are met (e.g., the sender holds the private key of the account, the address has enough Bitcoin to cover the transaction and fees, etc.). This is all executed automatically by code.

The Ethereum platform builds on this framework by allowing user-generated executable code to run on the blockchain so that parties can create their own smart contracts. These can be highly sophisticated contracts that go far beyond the simple settlement of payments. The blockchain can be used to automatically transfer records or digital goods upon payment. Many contracts can be moved entirely on blockchain. The blockchain can serve as an immutable record of truth not only about payment but also about ownership.

The real estate industry is beginning to experiment with this model. Imbrex runs the world's first distributed real estate platform, which is built atop the Ethereum network. Lemonade, a New York–based rental insurance company, automates contracts and the settlement of claims on blockchain. Contracts are offered, signed, and confirmed on the blockchain. The whole contract is executed and maintained by all parties. More and more companies are exploring how blockchain can improve the way they do business. Smart contracts allow companies to transact and interact directly with customers, vendors, and other stakeholders without the need for administrative middlemen.

The ability to execute distribution of digital goods and rights management on blockchain makes it ideal for anyone selling creative or intellectual properties. Content creators will benefit from doing away with the need for management or administrators. Consumers of their content will also benefit. Cutting out middlemen is a tried-and-true strategy for bringing down costs.

In the music industry, new musical acts have an easier time getting their music to listeners than ever before. The internet has democratized access to information and made it easy to get content to consumers. New bands no longer have to slug it out in the local scene until being discovered by a major label that can get them on the radio. Unsigned pop artists rise to fame on YouTube. New bands break out on Spotify. Rappers make it big on SoundCloud.

Nonetheless, big musical acts still sign with record labels that take a large cut of the proceeds. The record labels no longer even have to look for talented and exciting new artists. New artists do the hard work of building a large fanbase and only then do the major labels show up for their cut.

Major labels have retained power for a number of reasons. They still control much of the musical ecosystem. They control who gets on the radio and into festivals. Artists can go it alone through crowdfunding sites, and some have done so to great success. In 2012, Amanda Palmer famously launched one of the first super successful Kickstarter projects when she raised $1 million to record her next album. However, with more musicians turning to these platforms, raising substantial sums is difficult. Most acts are lucky just to get recording and production costs covered. And while getting your music onto Bandcamp, Spotify, or YouTube is easy, these platforms pay little in royalties. For this reason, most acts still sign with a label.

Many musical acts would do better if they could control distribution on their own. This would save them the cost of having to fork over most of the money from sales to various intermediaries. Blockchain could make this possible by creating true peer-to-peer platforms on which artists could distribute their product, accept payments, and manage intellectual rights without the need for middlemen. Distribution, payment, royalties, and even streaming could all be managed automatically on the blockchain. Streaming over blockchain would also make it easier for artists to protect their intellectual property by restricting who has access.

On a true peer-to-peer network, no middlemen would be required. An open-source blockchain could handle millions or billions of microtransactions over a peer-to-peer network. The industry could be decentralized and break the stranglehold that large conglomerates like iHeartMedia have on it.

ADMINISTRATION OF PUBLIC SERVICES

Blockchain has uses beyond the business world. Smart contracts can help better implement the social contract. Nowhere is this better exemplified than Estonia. After regaining its sovereignty from Soviet occupation in 1991, Estonia embarked on the ambitious e-Estonia initiative to become the world's first digital nation.

By all measures, they have succeeded. Estonia now offers 99 percent of government services online. Citizens can vote and execute virtually all other public services electronically. Estonia was implementing government services with blockchain-like technologies before blockchain even existed. Today, citizens can interact with all government agencies with the privacy and security of a public system built around blockchain architecture. In 2014, Estonia even implemented an e-residency program that offers digital residency—discrete from and not connected to physical residency—so that anyone in the world can participate in the e-Estonia project. The program is a first step toward a digital world without borders.

Blockchain is the ideal platform for many, if not all, government e-services. Security and privacy are important when dealing with sensitive personal information. Private keys could replace public identifiers, such as Social Security numbers, which would help combat fraud and identity theft. Smart contracts could be used to administer public services, welfare programs, and tax collection.

Voting is perhaps an ideal use case for blockchain. We care a great deal about maintaining election integrity. The efficiency hit of running elections on blockchain is well worth the added security and privacy. Citizens could vote from home while lowering the risk of voter fraud. Blockchain would make voting both more private and more transparent. Private keys would allow voters to ensure that their vote was counted accurately. The blockchain would forever exist as an immutable record of ballots. Elections officials could ensure that votes were counted accurately without exposing the personal details of the voters.

Organizations and companies should seek out other such use cases in which blockchain's qualities make it the ideal solution.

The Universal Worldwide Blockchain?

When author Darcy DiNucci coined the term *Web 2.0* in 1999 to describe the rise of user-generated content and greater interoperability across the web, various technologists and marketers ran with it. Every couple of years, marketing firms start pushing the nebulous concept of a Web 3.0 once again. However, Tim Berners-Lee, credited as the creator of the World Wide Web, has always seen the internet as a collaborative global project. Early blogs and websites featured user-generated content. Even the Usenet newsgroups and bulletin boards that predated the World Wide Web were a place for distributed users to come together to share ideas and information.

In other words, the peer-to-peer ethos of Web 2.0 wasn't really something new. What changed was just the result of wider adoption as the internet became more integrated and accessible. Practically everyone is now online. The companies dominating the internet for the last twenty years employ peer-to-peer business models, not the traditional business-to-consumer models of traditional companies, that harness the creative content of this expanding user base. The most visited websites are those built around user-generated content. Facebook, Instagram, Snapchat, Reddit, YouTube, Twitch, and similar sites make up the majority of web traffic. Wikipedia, which is maintained by users, is now the most used reference tool in the world. Craigslist captured the personals. Etsy is the new community art studio. eBay is the world's distributed garage sale. Even Amazon, despite being a centralized online retailer, is also a platform and fulfillment center for a network of sellers distributed around the world.

Although certainly easier to traverse now, the internet still has the same basic architecture it has had since inception. However, that may well change with the exponential growth of blockchain. Much, but not all, of the internet has moved onto decentralized peer-to-peer networks. There are understandable reservations about security and privacy that have kept certain parts of the internet from decentralizing onto peer-to-peer networks.

This is especially true when we look at the architecture of the internet. The large social media companies that generate and distribute user-generated content still do so from their own centralized servers. Theoretically, peer-to-peer networks could allow users to share generated content directly without the need for these intermediaries. Historically, this hasn't happened, due to concerns over privacy and security. However, blockchain can address these issues by allowing users greater control over their data and identity.

Currently, blockchain is computationally intensive. Deciding to employ blockchain architecture is often a decision about trade-offs. Added security and privacy have to be weighed against the efficiency hit of using blockchain. However, as the technology improves exponentially along with global computational power, we will one day be able to put anything and everything on blockchain. Given that blockchains can make discrete systems interoperable, we could actually put the entire internet on blockchain and decentralize the whole thing.

Why would we want to do this?

In a word: *security*.

The internet was never designed around security. The internet was created primarily for ease of access and democratization of information. The project has been a resounding success on those measures, but its underlying infrastructure is neither secure nor controlled. Fraud and online scams are rampant. Hackers hijack accounts, computers, and webcams. Computer viruses and ransomware are everywhere. Fake news is rampant. The back alleys of the internet are full of "deep fake" videos that use AI to render near flawless video and audio of specific people, often for nefarious purposes.

Addressing these problems has historically been a matter of applying security systems and measures overtop an inherently insecure architecture. We use passwords and firewalls and antivirus software and other security measures. We attempt to root out disinformation manually by fact-checking. These are ad hoc security measures meant to secure an insecure architecture. They are the equivalent of digging a moat around your house rather than building an actual castle for defense.

Blockchain would allow us to redesign the internet around both information sharing *and* privacy/security, not just the former. Security could be built into the internet's very architecture just as openness is now.

Someday we will need to redesign the internet—that's inevitable. The infrastructure was cobbled together and bandaged as needed. At this point, the internet is bandages all the way down. Someday we will have the computational power to move the entire internet onto block-chains and tie them together with a universal blockchain. This would allow the entire internet to be run as one big peer-to-peer network. Such a redesign would allow us to authenticate information as easily as we share it now. Fake images, videos, and news could be identified easily. Authorship could be verified by the system itself. Hackers and scammers could be more easily identified as unauthorized users.

That isn't our present, but with exponential technologies growing increasingly rapidly, it might well be in our future. When the day does come, we will have an internet that can *truly* be described as Web 2.0 in every sense.

In the meantime, organizations can employ blockchain where appropriate on a smaller scale. Blockchain projects can be used to improve specific business processes. Industries could create universal industry blockchains to make disparate systems more secure and interoperable.

Let's now look at what these applications of blockchain could do for the healthcare industry.

CHAPTER 8

BLOCKCHAIN IN HEALTHCARE

COMMUNICATION BETWEEN THE complex web of stakeholders that make up the US healthcare system creates a massive administrative burden. Providers, healthcare institutions, insurance companies, pharmaceutical companies, state and federal governments, and many other entities all have to interact with each other and the public. These interactions occur as a string of business processes. Even simple interactions have to be initiated, processed, and logged. Transactions must be created, submitted, and settled. Careful records must be kept for reference and to harvest insights. Executing these tasks and managing the ceaseless flow of information efficiently are key to bringing down healthcare costs.

Many of these interactions are automated, especially those that occur within discrete systems. Hospitals and even whole hospital systems can simply hire consultants to come in and implement automation. I did this work myself for years. Little outside collaboration is required to get an internal system automated.

However, hospital systems often need to communicate with one another. They regularly communicate with insurance companies. Primary care doctors interact with specialists and labs that have their own systems. Automating interactions across these systems requires more coordination. Intersystem interactions require the systems to be interoperable. Automating these interactions optimally requires more than just collaboration and standardization. The automated exchange of sensitive medical and financial information requires special attention to privacy and security.

The unique features of blockchain can make healthcare information systems interoperable while enhancing privacy and security.

Blockchain for Improved Information Access and Sharing

The information sharing problem the healthcare industry faces is one of decentralization—but the problem is not decentralization itself. The immense administrative burden the industry shoulders is the price we pay for a decentralized healthcare system. The burden is real—but decentralization has its merits and should not be tossed out.

Decentralization allows for flexibility and nimble action. Healthcare organizations and companies can build and rebuild fit-to-purpose solutions. Processes can be changed quickly according to local needs. Private actors rapidly emerge to fill unmet niches. Institutions can retool as necessary without the need for a concerted national effort.

Although centralizing the healthcare industry would make the flow of information more efficient, it would not come without cost. Privacy and security would be compromised. Databases can be hacked. Large centralized databases mean bigger hacks. Hacking a hospital database might compromise thousands of people's information.

Hacking a centralized national information system could expose the entire country.

There is also the matter of who would manage centralized servers. Would anyone actually want to shoulder the liability that comes with such a job? If so, why? Can we really trust the government or a monopolistic private entity with so much data? What would keep costs down if a central mediator were operating as a monopoly?

Centralization would also do away with the strengths of our decentralized healthcare model in exchange for efficiency. This is a mistake. We should instead find ways to make information access and sharing across a decentralized system more efficient. Toss the bathwater; keep the baby.

Historically, intersystem interactions have been the primary stumbling block for sharing data efficiently. Institutions maintain their own centralized systems. However, these systems need an efficient, orderly, and secure way to communicate, share distributed records, and settle transactions. Moving medical records and intersystem transactions onto a shared industry-wide blockchain would reduce friction in the flow of information without sacrificing decentralization and its benefits. Blockchain, as a distributed data structure, is a natural fit for our decentralized healthcare system. Medical records and data could stay distributed while still being accessible on the blockchain.

> Moving medical records and intersystem transactions onto a shared industry-wide blockchain would reduce friction in the flow of information without sacrificing decentralization and its benefits.

Such a blockchain would make the many information systems across the industry more interoperable. Individual systems would not have to change much to make this work. Blockchains are distributed ledgers maintained by participants. Medical records that reside at various hospitals, doctors' offices, and labs could remain in place while being made accessible on the blockchain. Hospital records stay with the hospital. Insurance records stay with the insurance company. Patient records can stay with the providers, institutions, and entities that produced and stored them or with the patient in some cases—whatever makes the most sense. However, with a little standardization, all healthcare-related records could be accessed by authorized parties via an industry blockchain.

Algorithmic keys would ensure that only authorized users could access specific records and only for an authorized window. This would allow various healthcare entities to manage the flow of information without duplicating and disseminating copies of records. Relevant parties communicate and transact directly by granting access to information on the blockchain. The doctor can release records directly to the patient. The insurance company communicates directly with the doctor. The insurance company interacts directly with the policyholder. These complex relationships are reduced to sets of two-way direct interactions *in an orderly, controlled, and efficient manner.* Authorized parties can access records distributed across several systems without jeopardizing their privacy or security. Records would also be less susceptible to data rot, corruption, or missing information from being copied and moved around.

This approach will forgo the security risks of centralization. No centralized servers or databases are needed. The industry would store and transfer data collaboratively over a peer-to-peer network. We are, in effect, centralizing *access* to health information without the need to centralize the storage of records. The records remain in place or wherever is most appropriate, while still being accessible to authorized entities. This removes the need for information to be stored in one place that could be easily compromised.

If all medical records are moved onto an industry-wide block-chain, healthcare consumers, providers, insurers, medical institu-tions, and other relevant parties can work from the same trusted set of records. Network participants would authenticate their identity and verify the identity of others with algorithmic keys. This allows for trust over a network of actors, even among strangers. Identity can be established and verified with near 100 percent certainty, allowing for the secure sharing of private information that has been authenticated by the network.

Standardization would allow AI systems to interact with records on the blockchain. Currently, records often have to be requested by phone, email, or fax. Someone on the other end has to verify the identity of the requester and authorize the transaction before releasing records. Blockchain would allow such processes to be fully automated. Authentication could be done by algorithmic keys and access granted via smart contracts.

Blockchain coupled with AI technology has the potential to make much of the administrative services industry obsolete. If every interaction is added to an industry blockchain, access and processing of information can be automated securely. Information can be made available to authorized parties instantaneously and automatically. Blockchain provides us with a secure way to automatically validate identity and grant access to records stored in various systems. Discrete automated information systems could trust the blockchain to serve as a shared record of truth. Interactions, settlements, and other business could be conducted with less friction and greater automation, all at reduced cost.

However, managing the flow of information more efficiently isn't just good for the bottom line—it's also good for the quality of care. Faster, automated transfer of records cuts down on wait times. Appointments could be made faster. Surgeries could be scheduled more quickly. Records could be released instantaneously.

Care is safer and more effective when information is readily accessible. Doctors and triage nurses have little knowledge of the medical history of patients who show up at the emergency room. The patient may not be fully conscious or ambulatory. They might have allergies or conditions that render standard care approaches inappropriate for the particular individual. This information isn't readily available to emergency responders or ER staff. With these records made available on blockchain, first responders could review records while the ambulance is on the way. ER doctors could have reviewed a patient's history before they even reach the hospital. This would be difficult to automate securely without blockchain.

The improvements to care aren't limited to emergency situations. Blockchain allows for easy access to a *holistic* record of a patient's medical history. Doctors often rerun labs that have already been run elsewhere when the information isn't on hand. They may not be aware that it already exists. This creates a lot of waste and subjects patients to more diagnostic radiation and needle pricks than are necessary. If these records were tied to a patient's identity on the blockchain, patients could easily release their medical history to their doctors. Fewer labs would be repeated unnecessarily. This would speed up care (no use waiting on lab results you already have) and reduce costs.

The gains of having a holistic, longitudinal view of a patient's medical history on hand cannot be overstated. This will improve case outcomes and reduce costs across the entire healthcare system. Easy access to better information will revolutionize care. Patients benefit from this more than anyone. Doctors can look back over the blockchain to pull a patient's entire medical history. On a population level, medical researchers could draw insights from holistic views of patient data to see what therapies work best in different contexts.

Consider how such a system might benefit someone like our hypothetical John.

Imagine John getting into a nasty car accident on the way to work. John proceeds through an intersection on green when another car runs the stoplight and T-bones him. The accident is bad, but the airbag deploys, and John was wearing his seatbelt. The car is totaled. But he's not dead, not yet. EMTs arrive on the scene and help him from the wreckage and into an ambulance.

At the emergency room, John is treated for broken ribs and a neck injury. He spends two nights in the hospital before being discharged. Afterward, he does ten days of outpatient physical rehabilitation.

None of this is cheap. Thankfully, John is well insured through his employer. The hospital stay is covered under his health insurance. The physical therapy is covered under his automobile policy. His health insurance company might have stepped in for that as well, but it didn't because he has other coverage that takes precedence.

However, for these claims to actually get paid, the different insurance companies need to work things out. John has medical insurance. However, the other driver was at fault. The insurance companies have to work out which owes what to avoid court. This involves a laborious document discovery phase. Personal statements and police records must be collected and compared against each company's coverage guidelines to determine who pays what. Records from the hospital and rehabilitation center are required. The facts of a big insurance case can be complex and take some time to resolve. A case like John's might take months to settle. Any unforeseen administrative problems could cause a lapse of care. Providers are hesitant to render services for which they may not be reimbursed.

The process could be made much easier if all of the information is codified on a universal healthcare blockchain as it happens. Guidelines should go on the blockchain as they are issued and updated. Medical records go on the blockchain as they are created. Even the accident information could be added to the blockchain as the police report is filed. These records could be standardized for easy machine reading. Then the entire claims process could be automated. Payments could be settled immediately so that there is no risk of interruption to John's recovery plan.

Blockchain-Enabled Identity Management

Historically, healthcare consumers have had little control over their personal medical records. Medical records are distributed across many systems and offices. Individuals have little to no say in how they are used. We have no way of knowing how our information is stored. Is it secure? Who has accessed it and why? Are there shared copies languishing in third-party storage, and how securely are those parties treating our data? We cannot know because we have no mechanism for controlling access to our own records.

This is a real problem given the sensitive nature of the information. Social Security numbers. Credit card information. Private information about medical conditions that we might not want to be made public to employers or anyone else. With this information scattered across dozens, if not hundreds, of places, the security risk is immense and snowballs over time. Security and privacy risks grow exponentially each time information is shared. The information is now stored in yet another location, maybe several, that can now also be compromised.

Ironically, although consumers have to worry about others accessing their medical information, they *themselves* do not have reliable access to all of their own records. There is no centralized repository or database for medical records. They are generally maintained and stored by the providers and institutions that created them. There is no easy way to compile these records. Patients cannot provide their whole medical history to new doctors. Getting medical records requires chasing them down from the various entities that hold them.

This is no easy task. Getting records often requires navigating various information systems and patient portals that use different identifiers for tracking and authenticating identity. You have to keep track of a dozen usernames and passwords. You will be issued various member IDs and group numbers. Hospitals have their own internal identifiers, often just your Social Security number, which is a security risk unto itself.

Requesting older records is even more difficult. The providers that have them may have moved or quit practice. Even if you can find the providers, there is no guarantee that they will be able to find what you are looking for. Many records are never even seen by the patient. How can you find or manage what you don't even know exists?

Some patients will find navigating these systems on their own

> Moving medical records onto blockchain and tying them to an industry-wide patient identifier— a set of public and private keys—would allow individuals to *exercise ownership over their own identity.*

too hard and simply give up. It's often not even worth trying to track down old doctors or specialists you have seen only once. Just reordering labs and exams, though wasteful, is often more practical.

Blockchain can address these privacy, security, and access issues by allowing streamlined identity management. Blockchain provides a secure way for entities to access and control records that are distributed across the healthcare system. Moving medical records onto blockchain and tying them to an industry-wide patient identifier—a set of public and private keys—would allow individuals to *exercise ownership over their own identity*. These keys would serve as personal identifiers in much the same way as Bitcoin private keys allow ownership of Bitcoin wallets. This could let consumers exercise control over records stored across a distributed network of healthcare providers and institutions. Those records could then be made available on the blockchain by the individual. The individual could also grant institutions the ability to share certain records with others.

In this way, medical records could be quickly accessed and controlled by the patient. Patients would also have the freedom to lock down their medical records or preauthorize access to certain institutions. Smart contracts could automate authorizations so that they could be issued or revoked automatically or manually. Blockchain would thus make patient records more secure while also being more accessible.

Someone like John could view his entire medical history instantly on any networked device by simply authenticating his identity and pulling the information from the blockchain. He has access to his entire medical history, which is tied to the private key that he alone holds. This information could then be shared with his doctor. John is in charge. Everything is encrypted and under his control. Only the information that John wants to share is shared.

This doesn't place any additional burden on John. The records stay right where they are stored with his various providers and are simply accessed on the blockchain. He doesn't need to chase records down and prove his identity over the phone. He doesn't need to worry about managing or securing that information on his own. But it's all right at his fingertips when he does need access.

This has all kinds of uses. John can release records to doctors when he visits. He can enter into a smart contract with first responders so that they can be granted access to his records in the event of an emergency, even if he is unconscious and unresponsive. This would alert EMTs about his high blood pressure and his penicillin allergy. This is lifesaving information—but only if they have access to it. Blockchain can help ensure that they will.

Data Monetization

Your health history is unique, like a fingerprint, but it can provide insights into traits and conditions when compared with medical records from the entire public. Researchers use this data to determine what behaviors and therapies induce or mitigate risk in specific populations. Your medical data is therefore valuable.

Healthcare institutions are aware of this value, which is why they collect patient data. They use the data to improve internal processes and therapies. Sometimes the data is sold on to third parties, such as pharmaceutical companies. However, healthcare consumers are never compensated for their data. Without any way to exercise control over personal medical records, individuals have little choice in how their data is used. The value of their unique medical information is captured by someone else.

Blockchain could fundamentally alter this dynamic by allowing consumers control over their own medical records. Tying medical records to a patients' private keys would allow them to control the data. No one could pull their records without permission. Third parties that want to monetize data would need to get consent. Consumers could therefore capture the value of their own health information by selling it directly to interested parties, if they so wanted. Some might gift the data. Others might opt out entirely.

This isn't just a win for patients—it's also a win for the healthcare industry at large. Blockchain would make it cheaper and easier to access consumer data by purchasing it directly from patients. This is a good thing. The entities buying medical data do not usually have nefarious purposes. The demand for medical data comes mostly from within the healthcare industry. Hospitals use patient data to design better systems and processes. Medical researchers and drug companies use the data to develop new and better therapies. Patient data is critical to research and development. Putting that data into the hands of researchers benefits all of society. Blockchain makes getting patient data into the hands of those who need it more efficient and thus less costly. There isn't any need for middlemen. The consumer can sell the data directly to researchers, which would lower costs.

As we saw in previous chapters, the process of taking a new drug into development, through clinical trials, and all the way to FDA approval is extremely expensive. A new drug generally costs millions or even billions of dollars to bring to market. Although there are many inefficiencies driving up the cost of pharmaceutical research and development, one of them is the cost of acquiring participants in clinical trials. Each trial has unique needs, which require participants with specific characteristics. Trials for a new leukemia drug targeting a specific blood type require leukemia patients with the right blood type to participate. The subjects may also need to have been diagnosed within a certain time frame. Comorbidities might exclude participants. They might need to come from a specific age range or BMI category. They might need to be balanced by sex or some other characteristic. Participants in clinical trials might need to meet all kinds of criteria, which is why drugmakers pay clinical research organizations to recruit subjects.

These recruitment services drive up costs. Blockchain would allow this service to be automated. People could opt into being study participants. Automated systems could scan the blockchain for participants who match the specific criteria of individual studies and alert them automatically. Blockchain would allow this to be handled privately and securely. Records don't need to be centralized and identities don't need to be revealed for this kind of analysis and transaction to be handled. Blockchain allows for trustless direct interactions between the buyer and the seller while keeping the information private and secured. Participants could thus keep their identity private until entering into a study. Researchers who just need your data wouldn't ever have to know who you are.

Such a system would allow researchers to find study participants nearly instantaneously without having to pay recruiters. This would ultimately help bring down the cost of running clinical trials and of developing new therapies. Individuals would also be able to offset some of their own healthcare costs by monetizing their own data.

Blockchain-Assisted Virtual Care

In previous chapters, we explored the nation's dire need for a more robust virtual care system. The COVID-19 pandemic has laid bare the high cost of underinvestment in telehealth. Forcing patients into direct contact with the healthcare system for services and appointments that could be offered online puts them and providers at unnecessary risk. Controlling transmissible disease comes down to breaking chains of transmission. Keeping sick people at home when possible protects everyone, including those the patient might come into contact with on the way to the doctor.

This was true before the pandemic, and it will be true afterward as well. Hospital-acquired infections were a serious problem long before COVID-19. Hospitals are notorious for this problem. Expanding virtual care to keep people out of hospitals and other medical facilities as much as possible will reduce disease and lower overall healthcare costs.

Quality need not be sacrificed. Telehealth isn't an inferior substitution for in-person service—it is often preferable. Telehealth is often more efficient and less expensive than offering the same services in person. Many consultations and evaluations can be conducted over video chat. Prescriptions can be ordered online, and many areas now enjoy same-day delivery. Many labs can be done from home and mailed in. We should be fully leveraging these conveniences to deliver healthcare more efficiently.

Of course, many healthcare services can only be performed in person. You probably won't have a cast set or a tumor excised from the comfort of your living room sofa. However, post-op care for these procedures often can be conducted virtually.

Blockchain won't be the technology that delivers virtual care. Virtual care will rely on video conferencing and patient portal systems. Those with chronic conditions might use in-home sensors for monitoring certain vitals. We already have technologies for delivering telehealth—we just need to use them. However, blockchain could facilitate the efficient and secure delivery of virtual care.

Telehealth is highly compatible with blockchain technologies. Telehealth is already delivered online. Records are necessarily digital. That information could be automatically added to the blockchain. The information already has to be plugged into the system. Why not secure it and make it accessible and manageable via blockchain?

Patient records and treatment plans could be shared with relevant parties immediately via blockchain. Specialists could immediately access treatment plans. The insurance company would immediately see that referrals had been made. Everyone would remain in the loop by working from the same verified, holistic record of truth about each medical case. Everyone gets the information they need—and only the information they need—in a secure manner. With standardization and automation, much of this could be done effortlessly and instantaneously.

Telehealth isn't necessary to deliver care. Blockchain isn't necessary to build a virtual care system. But virtual care makes healthcare more efficient, and blockchain can make virtual care more effective. Digital records would be produced and made available immediately. Everything would be automated and secured. The patient wouldn't have to rely on doctors and administrators to collect records—the patient could make them available themselves. All of this can be done without the security risks of centralization or the burden and added cost of administrators.

Virtual care is necessary to bring down healthcare costs. We will need to overhaul the healthcare system and our infosystems to make that happen. We might as well implement and integrate blockchain where possible to make new systems more secure and resilient.

Blockchain and Elder Care

A sad fact of life is that many people lose the ability to manage their own healthcare as they age. Failing health makes it harder to take care of yourself when you need to the most. Children often have to step up and help care for aging parents. This can involve great sacrifice, especially when adult children live far from their aging parents. The children of aging parents who can no longer care for themselves are frequently faced with difficult decisions. Many want to care for their parents but simply cannot walk away from careers, mortgages, and friends and family. People too often have to choose between their parents and their own lives.

Blockchain has the potential to make managing care for the elderly much easier, even from afar. Elderly people with failing health could authorize a child or other trusted caregiver to make certain decisions on their behalf. This permission could be granted on blockchain and executed via smart contract. The parent doesn't even have to sign over their passwords or accounts. They simply authorize a third party to manage defined parts of their healthcare.

This would make it much easier for someone else to help manage your care from afar. The trusted caregiver could pull records and follow up on appointments easily. They could interact with providers. Authorizations can be issued and verified easily on the blockchain. Blockchain makes this simple and secure. Providers would have verified proof that the person making decisions on your behalf is authorized to do so. This would improve elder care greatly. Providers are often flying blind when working with disabled or elderly patients who cannot manage their own care.

Blockchain would also put more control back in the hands of the elderly patient. Smart contracts allow us to make decisions about our end-of-life or late-life care while we are still capable of doing so. Those contracts could be preserved indefinitely on the blockchain. Smart contracts would also allow for more privacy around end-of-life decisions and the rest of our personal medical affairs. Caregivers could be authorized to share records without actually being granted access to the information themselves.

Consider how these advantages might improve life for someone like our hypothetical John.

Fast-forward several decades. John is older now. His father has passed away. His mother just turned eighty. John is still working in Silicon Valley. His mom lives on the East Coast. When she slips in the shower and fractures her hip, both of their lives become more difficult. His mom is rushed to the emergency room. She doesn't need a total hip replacement, thank God, but she does need screws put in to stabilize the fracture. The procedure goes well and is done before John can even get to the East Coast.

Things get more complicated from here. Surgery is quick. Recovery takes time. His mother needs weeks of daily rehab followed by months of outpatient physical therapy. Seeing a variety of specialists is key to her continued recovery. Unfortunately, the early signs of dementia make it impossible for her to manage her own care. John has to do it for her from afar. Having just moved into a leadership position with his company, he is not at a point in his career where he can just walk away. Like so many middle-aged Americans, John finds himself sandwiched between kids and aging parents who both depend on him. His mother needs help getting to her appointments, but his kids need a college fund. He cannot just abandon his job to be with his mother.

John tries to manage his mother's care from the other side of the country. Suddenly he finds himself always on the phone with doctors and Medicare agents. He squeezes calls in early in the morning before heading to the office, while on break, or in the evening before heading home—whenever he has a moment and can get through on the line. There is always some medical appointment to schedule or follow up on. He has to arrange for transportation to and from physical therapy. He has to find someone to ensure that she is taking her prescriptions properly. John is always worrying that some new doctor won't be warned about her drug allergies. He worries that she won't get the support she needs as someone in cognitive decline. John tears himself apart trying to prevent crises before they happen while still managing work and family life. It's not easy, but what else can he do?

Now imagine that John lives in an America with a healthcare system that has invested heavily into virtual care systems, automation, and blockchain. John would have become an authorized primary caregiver for his mother so that he can manage her care from the blockchain. He would have access to all of her records, which would be updated in real time, so that he could ensure that her providers are up to date. Immediate access to updates in her treatment plan would also keep him in the loop. He might join some telehealth appointments with her.

Blockchain would allow John to work with his mother's providers with the same playbook and game plan in hand. Each update or change to her treatment plan would be codified on the blockchain. John would always be informed and in control. So would his mother's providers. John would always know what services his mother needs and is getting, whether she made it to her appointment, and everything else. He would be able to work directly with her doctors in an informed manner.

Best of all, none of this requires much of John or anyone else. He doesn't have to spend hours on the phone chasing down doctors. He doesn't have to hound the insurance company or the pharmacy or the rehabilitation center. All of the necessary information is available on the blockchain. John could rest easy from the other side of the country knowing that his mother is getting the care she needs.

These are both hypotheticals. One of them will be our future. We get to choose which. It's a matter of where, as an industry and society, we choose to make investments for tomorrow.

CHAPTER 9

ALIGNING INCENTIVES ACROSS THE HEALTHCARE INDUSTRY

THIS BOOK HAS BEEN an exploration of the problems facing the American healthcare system as well as the solutions offered by novel exponential technologies. Managing the inefficiencies of decentralization without also forgoing its benefits hasn't always been possible. But it is now. Maturing exponential technologies—artificial intelligence and blockchain, in particular—promise to address systemic inefficiencies in the American healthcare system while offering greater privacy, security, and effectiveness, all at lower cost.

Sound too good to be true? It's not. We really can build an efficient decentralized healthcare system. Automated processes and data analysis, AI decision-making, blockchain-based information systems, expanded virtual care, and insight-driven preventative care will revolutionize what is already the best healthcare system in the world.

Skeptics may ask, if these exponential technologies are so great, why aren't we already pursuing them? Wouldn't the market have restructured the healthcare system around these technologies if they hold so much potential?

Well, the industry *is* deploying exponential technologies to improve care and lower costs. Automation is common throughout the industry. AI already handles much of the repetitive decision-making, rote data collection, and quality-improvement analytics. Better data-driven preventative care is recognized as crucial to controlling costs. Virtual care is expanding, especially in the wake of COVID-19. Cutting-edge healthcare companies, such as Anthem, are now exploring blockchain solutions to various business problems. The industry is also deploying other exponential technologies, including advancements in data science and robotics in several areas. These technologies *are* being rolled out in new and exciting ways.

However, we aren't deploying them fully or rapidly enough. This is true. The reason, as we touched on earlier in this book, is that the healthcare industry's many stakeholders are all pulling in different directions. Everyone has different incentives, and those incentives are often not aligned. Providers want one thing, insurers another. We are underinvesting in preventative care because reactive care is more immediate and profitable. Insurers cannot be bothered to make forward-looking investments in policyholders who may not stick around. Long-term interests in public health and economic sustainability are eclipsed by short-term profit motives.

Meanwhile, patients don't comply with doctor's orders or look out for their own best interests. The long-term consequences of daily actions are misaligned with what people want in the moment. Patient compliance with preventative care guidelines is often poor.

The whole industry is mired in a collective action problem. What is best for the healthcare system as a whole, and for its constituent stakeholders *if* everyone acted collaboratively, is often not what makes sense for individual actors operating under the status quo, especially not in the short term. The result is a healthcare system that costs more and delivers less than ideal health outcomes.

Addressing misaligned incentives is key to transforming the healthcare system and bringing costs back down to earth. The technological solutions to our problems exist, but solutions are worthless unless implemented. The whole industry, as well as the public it serves, must come together to make the necessary changes. That won't happen unless we change the incentive structure that drives conflicting behavior and suboptimal outcomes. The industry isn't going to change unless its stakeholders are given a reason to do so, and neither will the public. That's just basic market logic. It's basic human nature. Providers, healthcare institutions, insurers, regulators and policymakers, and society have to be given reasons to make changes. Investing in novel technologies and redesigning systems around them require up-front investment in capital, effort, time, and attention. Healthy lifestyle choices require people to put the future ahead of the present moment. Those investments and better personal choices won't be made without having the right incentive structures in place.

> **Ultimately, the choice to transform the healthcare system is just that—a *choice*. It's a policy choice. But the right policy and behavioral choices must be made collaboratively as an industry and as a society.**

Ultimately, the choice to transform the healthcare system is just that—a *choice*. It's a policy choice. But the right policy and behavioral choices must be made collaboratively as an industry and as a society.

So far, we have failed to do so.

But what if the very exponential technologies that could transform the healthcare industry could also be leveraged to facilitate their adoption? What if these technologies could be used to set policy that would realign incentives across the healthcare sector so that data-driven preventative care is the default and not an add-on?

They can—here's how.

Resetting Industry Incentives

Providers, hospitals, insurers, researchers, drug companies, and other healthcare institutions are *market actors*. They serve a public good. Many of them got into health and medicine for altruistic reasons. But they still operate within an industry and respond to market forces. They won't invest in new technology, change treatment protocols, or expand preventative care without sufficient reason.

Market incentives drive providers and medical institutions to prioritize reactive care. Performing procedures and services is more lucrative than preventative care, which often comes down to simple lifestyle changes and regular screenings. Preventative care, by its very nature, prevents disease. That's great for public health and reducing the financial burden that healthcare places on individuals and society. However, preventative care lowers the need for healthcare providers and their services. A healthier population needs fewer therapies, surgeries, drugs, and all of the other costs associated with chronic health conditions.

Doctors and hospitals aren't conspiring to keep people sick. They simply treat sick people as they come through the door and have no real incentive to focus on preventative care. Providers want the best for their patients, but the people coming through the door are ill. People don't think about preventative care when they feel well, and by the time they are sick, it's too late for prevention.

Better data-driven insights into the *specific* benefits an *individual* patient will reap from preventative care would help make the benefits more real to both patient and provider. Personalized patient health reports based on AI-generated insights would help clarify these benefits. The industry should prioritize ongoing health reports with an emphasis on preventing chronic illness before it sets in.

What preventative care measures produce the best outcomes for specific people? When and where are they best applied? Different bodies present different risk profiles. Preteen girls have different risks than middle-aged men working construction. The more targeted our preventative care measures, the better our outcomes will be. People are often lumped into broad categories that may or may not share their risk profile in different areas. Doctors make care decisions by intuition, which can be faulty and based on stereotypes or pseudoscience. Various agencies and institutions issue guidelines, but these are not necessarily grounded in the best science either. Care guidelines don't always take advantage of all the data we have available.

Preventative care guidelines should be based on insights gathered from unbiased scientific analytics and tailored to the individual. Data from millions or billions of people can be safely analyzed over blockchain. AI systems can extract insights from the data in order to create *personalized* risk assessments for each patient. This takes the guesswork out of determining an individual's risk profile.

Strong insights into what preventative strategies work for specific individuals will go a long way toward convincing insurance companies to cover them at higher rates. Insurance companies have an inherent interest in reducing the claims they pay out. A clearer view of how preventative care can reduce a specific policyholder's total lifetime claims would encourage insurers to make those initial outlays. Higher reimbursement for preventative care would also further incentivize doctors to focus on these services.

Unfortunately, although preventative care reduces the lifetime healthcare cost of patients, the short-term transactional nature of insurance policies still disincentivizes such investments. Insurance companies don't see a return on investment for preventative care when policyholders switch insurers, as most do many times over the course of their lives. Preventing or postponing the onset of chronic disease can be a huge return on investment, but it often won't be realized for years or decades after the policyholder has moved to a different insurer. The return will likely not be realized by the insurer that has to make the initial investment, which is a big disincentive to investment.

> **Governments and regulators must create a framework that incentivizes insurers to promote and cover more preventative care.**

Overcoming this problem will require actual policy. Governments and regulators must create a framework that incentivizes insurers to promote and cover more preventative care. This can be done in several different ways.

The most direct route is simply having the government pay for preventative care directly. Creating such a universal benefit would be straightforward and easy to implement. However, this might introduce inefficiencies into the healthcare market. Insurers might be incentivized to cover preventative care with limited, marginal, or even no utility. In order to keep this from happening, policymakers would have to make nuanced decisions about healthcare policy that they are not always qualified to make.

A better strategy is for the government to create a framework within the market that incentivizes preventative care. Maintaining market incentives will help prevent misallocation of resources.

One such framework would be a system in which insurers are reimbursed for the cost of preventative care when a policyholder switches to a different insurer. The government could simply directly reimburse insurance companies for past preventative care spending when policyholders switch to another insurer. This could also be handled by the industry itself via blockchain with just a little bit of regulation. Preventative care credits could be logged on a blockchain and tied to a policyholder's identity. When the policyholder switches insurers, the new insurer automatically reimburses the old insurer for preventative care spending before the new policy becomes active.

Such a system would incentivize insurance companies to invest in preventative care without performing services with needless or marginal utility. Insurers won't hesitate to invest in effective preventative care when they know they will be reimbursed if the policyholder moves to a different insurer. This will promote effective—and only effective—preventative care and drive down costs for everyone.

Such a program would align the incentives of individual insurers with those of the wider industry. The insurance sector will then in turn properly incentivize providers and medical institutions to invest in effective preventative care. Health providers depend on insurance reimbursement, which gives the insurance industry enormous leverage over recommended treatment plans. Knowing that they will be reimbursed for cost-saving preventative care, insurers will actually promote such care to the benefit of all.

Realigning Patient Incentives

Of course, the individual healthcare consumer has a role to play here. Preventative care requires sustained patient compliance. This isn't as easy as getting someone to finish a course of antibiotics. Preventative care involves adopting healthier lifestyle choices *for life*. Unfortunately, what is best for long-term health isn't always what we want in the moment. Good health comes down to making the right choices now.

No one can force people to be healthy. However, as an industry and as a society, we can incentivize better health and lifestyle choices.

There is a public interest in doing so. Individuals have to live with the health consequences of their own choices. However, society also bears the consequences. When someone makes poor lifestyle choices that result in expensive chronic health problems, we all pay for the cost of their care. The insurance company may pick up the tab, but they always pass it right back to the policyholders in the form of higher premiums. The catch to collectivizing risk is that while other people take on your risk, you also take on theirs. Insurance allows us to pay for catastrophic health outcomes without going bankrupt or simply dying for lack of care, but the collectivization of risk disincentivizes personal responsibility for one's health. If you get sick, the insurance company will pay for it—no big deal!

Except it is a big deal—poor lifestyle choices and lack of preventative care are driving up insurance premiums for everyone and slowly bankrupting the country with runaway healthcare costs.

We need an insurance system to mitigate individual risk. However, to be financially sustainable, that system must contain incentive structures that encourage healthy lifestyle choices and a proactive approach to personal health. Historically, we have taken both a carrot and a stick approach when building these incentive structures.

Insurance companies offer discounts for not smoking, maintaining a healthy weight, and other healthy lifestyle choices while penalizing unhealthy choices. Many insurers offer vaccines at no cost. Children aren't allowed to enroll in public schools unless they are immunized.

These incentives work, but we could be leveraging technology to better target and promote them. AI-generated insights can reveal which incentives are most effective for whom. Incentives that are targeted at individuals and delivered on platforms that are integrated into their everyday lives will be far more effective.

Imagine a national patient portal that delivers personalized information on your most pertinent health risks. Presented in an engaging manner, such a system would incentivize users to make better choices by illustrating *exactly* how your daily choices affect your health and finances. Imagine engaging visualizations that present your own personal healthcare profile. Imagine being able to see your medical records aggregated and compared to those of people with similar traits and risk profiles. You would be able to see how your daily choices impact your long-term health. You could see the monetary incentives you are getting or missing out on based on your habits. Presented in a personalized and engaging manner, such a platform would make the connection between behavior and consequences clearer.

Being able to actually *see* the long-term consequences of your actions can be a powerful motivator. Consider the role mirrors play in grooming; they provide instant feedback on how we look. People do touch-ups when they pass by mirrors throughout the day. We straighten our clothes, touchup makeup, fix our hair. Mirrors remind us to pay attention to how we look. They remind us to attend to our appearance and help rectify issues. Similarly, daily health reports that provide continuous and instantaneous feedback on how our choices affect our health and finances provide daily reminders to do better.

Making good choices throughout the day is easier when we can see how our decisions affect us personally.

These daily health reports could be coupled with traditional carrot-and-stick methods of rewarding good choices and punishing bad ones, only in real time so that people can see the positive and negative effects of their choices. Tracking health markers and behaviors would allow such a system to reward people with immediate discounts for engaging in healthy behaviors. Exercise and weight loss could be rewarded. Sedentary behavior and unhealthy behaviors could be penalized.

This kind of targeted reward system would video gamify preventative care, tapping into the brain's reward structures in order to induce better choices. Simply offering a monthly discount for having a gym membership that you may not even use isn't particularly engaging or incentivizing. Watching your monthly insurance premiums increase automatically when you stop showing up at the gym is much more motivating. Even more motivating is seeing how that visit to the gym lowered your premiums.

Blockchain will be central to getting people to share their personal data on such a platform. Personalized insights require the personal data of individual users. The more data, the better. Integrating the platform into the greater economy in order to track consumer and lifestyle habits would make insights more powerful. The platform could track grocery shopping and food intake as well as physical activity, environmental factors, and other data to get a better picture of how the choices we make affect our health. However, people are understandably hesitant to share personal information. They worry about personal data being misused. Health is a private matter. People don't want their medical records falling into the wrong hands. They worry about identity theft and more general privacy concerns.

Blockchain will help overcome this trust barrier. Personal data doesn't have to be "collected." Analysis can be performed on the blockchain without sharing the underlying data. Properly implemented, such an architecture eliminates privacy concerns while still allowing insights to be gathered from all existing data.

Removing the trust barrier aligns the incentive of consumers with one another. They would be incentivized to make their data available on the blockchain in order to benefit from personalized health reports. Individuals could even monetize their data by selling it to the system. However, the greatest advantage is the insights themselves, which benefit all of society. Drug researchers could build better therapies. Providers could create personalized treatment plans. Insurance could develop better preventative care plans and evidence-based coverage guidelines. It's win-win-win all around.

The Virtuous Flywheel of Individual Choice

None of this will be possible without collaboration between the government, the public, and the many stakeholders of the healthcare industry. We all have to work together to create a better healthcare system that promotes better individual health choices. This country pays too much for healthcare because too many people are getting sick. The country is failing us. The healthcare industry is failing us. We are failing ourselves.

Turning the ship around will require us all to work together. We all have a stake in building a better healthcare industry and a healthier society. We won't get anywhere with everyone pulling in different directions. Regulators and legislators need to create a policy framework that aligns incentives of various industry stakeholders. Providers and medical institutions need to work together to produce better health outcomes by focusing on preventative care and cutting-edge health technology. Insurers need to work with medical institutions and pharmacy companies to deliver care at affordable prices. Individuals need to make better choices about their long-term health. The industry needs to make it easier for them to do so. Collaboration and doing our part is how we will make a good health system great.

> The individual choices we make build momentum with each turn of the flywheel. Eventually, if we are all turning together, the momentum becomes not just self-reinforcing but the very wind at our backs.

In his book *Good to Great*, Jim Collins argues that there is no one moment when a company moves from being merely good to *great*.[23] Instead, the process is akin to pushing a flywheel until the building momentum rockets the company to new heights not yet imagined. This is true of companies. It's also true of whole industries. And it is true of individuals. The individual choices we make build momentum with each turn of the flywheel. Eventually, if we are all turning together, the momentum becomes not just self-reinforcing but the very wind at our backs.

This book has focused heavily on how exponential technologies can help deliver better-quality healthcare more efficiently at lower cost. Automation, AI, blockchain, telehealth, medical insights, incentive platforms, and customized health reports—these are tools for delivering better healthcare more efficiently. They can also be tools for

---

23 Jim Collins, *Good to Great: Why Some Companies Make the Leap ... and Others Don't* (New York: HarperCollins, 2001).

creating the systems and platforms that can help us work together collaboratively. Technology can help make adversarial interactions more symbiotic, at every level, until we are all making the right choices together. I like to think of this as the *virtuous flywheel of individual choice*. We all have to take responsibility for our individual choices—providers, insurers, drugmakers, hospital administrators, and yes, individual policyholders and patients—but when we act in concert, the flywheel eventually starts turning on its own.

How might this look for the average patient and policyholder, someone like our hypothetical John?

Imagine a future where John wakes up each morning and checks his phone. Along with work emails, that early-morning text from his mom, the daily news, and his stock reports, John also receives a personalized daily health report and health industry news.

His daily health report generally has no surprises. They have been coming every day for years. Engaging visualizations allow John to actually *see* how his daily choices help manage chronic diseases and lower his personal health risks. This has allowed him to commit to managing his diabetes and protecting his health. A monetary incentive was particularly helpful in getting John to reduce his sugar intake.

These reports are custom tailored for him. His daily choices are recorded on the national healthcare blockchain so that insights can be gleaned from his data. Researchers, providers, hospitals, and other institutions use his data to better understand, prevent, and treat diabetes as well as other conditions John faces and is at risk for. Researchers use his data to develop better therapies and preventative care measures. His data helps create customized treatment plans for people similar to him, just as their data helps others. The small monetary incentives that helped John reduce his sugar intake were fine-tuned and offered to him because the insurance company knew that people like John would respond well to this incentive.

In little ways like this, billions and trillions of times over, the healthcare industry makes small tweaks to align the incentives of the industry and the public. Society has thus become much better at managing chronic health issues. Things are only getting better as we collect more data, mine it for more insights, and develop ever better therapies and technologies.

Meanwhile, state, local, and federal governments also monitor the national healthcare blockchain for anonymized data from all the Johns and Janes of the world. Governments partner with healthcare experts to build better policy and with industry to establish a policy framework for keeping incentives aligned.

Amazingly, and to John's gratification, he is paying less for healthcare now despite all the new benefits and services he receives. Preventative care and improved efficiency have brought premiums down for everyone. His insurance company rewards him with discounts for staying healthy and being proactive about his health. Every dollar that the insurance company invests into keeping John healthy is logged on the national healthcare blockchain so that his insurance company will recoup the cost should John ever switch insurers.

John also offsets some of his healthcare costs by making his data available for analysis on the blockchain. The money isn't a lot, but it allows him to monetize his health information and unique personal characteristics. John offers his data to certain diabetes researchers for free. He knows that his data helps people like him, just as their data benefits him. Besides, as a tech person, he understands better than anyone how blockchain technology keeps his personal information anonymous, safe, and secure.

John further offsets costs by participating in research studies and clinical trials. His daily health news comes with a curated list of studies for which he would be a good candidate. Participation often results in remuneration. John doesn't much care, though, as he still has his good job in Silicon Valley. The real reason he participates is because he understands that healthcare research benefits him and everyone else as well. He is healthier than he has ever been and wants the same for others.

John understands that nothing is more important than health. We all have a stake in building a healthier society. This is why he makes good individual health choices and it is also why he shares his data for analysis. This is also why his insurer now offers him incentives that help him meet his personal health goals, which are now aligned with the goals of his insurer, providers, healthcare network, and society at large. No one benefits from John getting sick and burdening the healthcare system—not John, not his insurer, not the country at large. Not anymore.

Such a future, and more, is possible for all of us Johns and Janes of the world. We have the technology. We just have to use it, together.

CHAPTER 10

THE TECHNOLOGY ADOPTION CURVE

REALIGNING INCENTIVES ACROSS the healthcare industry is no small task. The American healthcare system will have to be reimagined and rebuilt. We need to deploy new tech, new infrastructure, new protocols, and whole new ways of thinking about public health and our individual relationship to healthcare. No one person, one company, or one institution can make this transition unilaterally. We need a collective endeavor. The entire industry and all its stakeholders must get on board. Such a major societal change won't happen overnight. It won't be easy or without growing pains. But a radical transformation on a reasonable timeline is possible if we all take part.

And I do mean *all* of us. Leadership will need to develop and implement new strategies and solutions. Healthcare teams will have to learn new protocols and adjust to new systems. Healthcare consumers will have a critical role as well. Consumers are the primary drivers of markets. Their changing expectations will force organizations to adopt better technologies and digital solutions. As patients come to expect a better consumer experience, one more on par with other industries deploying new technologies more rapidly, healthcare organizations and teams will have to rise to the occasion. The adoption of new technologies follows a predictable curve within industries. Greater adoption enacts competitive market pressures that drive even greater adoption. The healthcare industry is currently lagging behind for reasons we will discuss next, but we are on the precipice of a great transformation.

While we all have a part to play in this transformation, some roles are bigger than others. Readers of this book are likely to be, like me, in leadership positions in the industry. I spent the last three years overseeing digital innovation at Anthem. People in a similar position have the special ability, as well as the duty, to directly shape the future of the healthcare system. Our performance will determine the shape of the healthcare system and the fate of our individual organizations.

Falling behind the Curve

In the last two decades, over half of Fortune 500 companies have disappeared in mergers, acquisitions, or business failures. Many of these lost companies were casualties of digital disruption. Things don't always have to end this way. Over fifty companies have remained on the list since 1955. Legacy companies can compete with tech-driven upstarts by deploying the same novel technologies and digital business strategies. You don't have to be a tech company to keep pace with the adoption of new technologies.

Different industries have different technology adoption curves, which is the rate at which new technologies and digital solutions are deployed. Falling behind the curve in your industry generally results in subpar performance and, ultimately, business failure. Some industries move faster than others. The tech industry obviously moves quickly. Their business is literally tech, which means that the space will naturally be full of innovators working with the latest technologies. Organizations in industries with lots of innovators have to deploy new technologies rapidly. Those that don't keep pace will fail. Organizations in industries with fewer innovators are much less subject to market pressure to keep up with new technologies. Technology laggards survive more easily when they are mostly only competing with each other. This creates a negative

The healthcare industry typically lags behind other industries in technology adoption by a full decade.

feedback loop on an industry's technology adoption curve. The fewer innovators and more laggards in an industry, the less pressure there is for the industry to ever catch up.

This is the unfortunate state in which the healthcare industry finds itself. The healthcare industry typically lags behind other industries in technology adoption by a full decade. Healthcare organizations remain viable without keeping up with deploying new technologies seen in more competitive industries. This is a lifeline to healthcare organizations that don't stay on the cutting edge of technology, but it comes at significant opportunity costs to society. This book has made the case for the power of novel exponential technologies to revolutionize the healthcare industry. The technologies exist now. The healthcare industry simply isn't exploiting their full potential. Healthcare organizations have little incentive to experiment with new technologies to find new and better solutions to our current problems.

This is not entirely without reason. The healthcare industry has a risk-averse culture rooted in the Hippocratic oath to *do no harm.*

"Move fast and break things" doesn't sound so great when your life or health are on the line. Clinical settings call for great caution when altering care protocols. Drugmakers and the FDA cannot roll out new drugs or other therapies without substantial testing. The high stakes of the healthcare industry require an abundance of caution that makes experimentation slower, more difficult, and ultimately costlier than in other industries. The result is a sclerotic industry hesitant to experiment with new digital and technological solutions for fear of risk.

This inertia isn't limited to clinical settings. The same cautious culture carries over into administrative offices, boardrooms, and IT departments throughout the industry. Healthcare organizations are generally slow to adopt new systems that might disrupt the flow of daily operations, whether they engage in direct services or not. The perceived risk is greater than the *also perceived* reward. This is because most organizations aren't factoring in the digital solutions they haven't yet discovered. Experimentation with these technologies would result in digital solutions that improve services and operations. However, without the market pressure to go looking for them, most healthcare organizations simply don't. With so many technological laggards in the industry, there is little pressure to experiment with new technologies that competitors aren't using either.

Consumers cannot act as an effective check on an industry falling behind with its technology adoption curve. Healthcare organizations aren't falling behind each other—they are collectively falling behind more nimble, innovative industries. Healthcare consumers have little choice in the matter because they can't vote with their wallets for enhanced services no one in their area is offering. Few providers and organizations are using new technologies to offer better services and experiences, so consumers have to be happy with what they get. It doesn't help that the healthcare market offers less of a novelty premium to early adopters than other consumer sectors, such as retail. Healthcare consumers are happy with services that are good enough, especially when they don't know what they could be getting. Healthcare organizations that aren't rewarded with more business for adopting new technologies simply don't offer them. The whole industry would perform better if they did, but there is nothing to kickstart innovation in an environment that doesn't encourage this competition.

Competition would force organizations to keep up with a steeper adoption curve. Innovators and early adopters accelerate the adoption of technology across whole industries. Conversely, laggards drag down the whole industry's rate of adoption. Consumers have little say in the matter with so few options available, and healthcare companies have little incentive to experiment with new solutions that bring risk with little reward.

Nothing about this situation is stable. First and foremost, there *are* innovative companies in the healthcare space. Anthem has made a commitment to being a digital-first company—and we aren't the only ones. However, we aren't doing this because of current market pressures. We are trying to create market pressure. The more companies that commit to being innovators, the greater the risk to others of being left behind. No company should want that for themselves. No industry should want that for itself either. And yet we have allowed the healthcare industry to lag behind in the adoption of technologies that would allow for better, more personalized healthcare services to be delivered more efficiently.

Black Swan Events and Hidden Systemic Risks

The healthcare industry falling behind the adoption curve presents a great opportunity cost to all of society. Healthcare in this country is more expensive, less efficient, and far less resilient than it needs to be. These opportunity costs go mostly unnoticed … until they don't. Healthcare executives do not only underestimate the perceived benefit of experimenting with new digital solutions—they also misperceive the risk of forgoing them. Black swan events can expose hidden systemic deficiencies that could and should have been addressed.

The coronavirus pandemic was one such event. COVID-19 exposed cracks in the healthcare system that had long gone ignored. The need for more surge capacity nearly brought the healthcare system to a breaking point. Hospitals were short on ICU beds, personal protective equipment, and trained medical personnel. Resources and personnel had to be diverted from routine and nonacute care in order to handle the surge in coronavirus cases. This situation was obviously not ideal. Routine care can itself be life or death. A postponed mammogram can mean that treatable tumors metastasize and become fatal. Regular screening and routine care often prevent chronic or life-threatening conditions from progressing in the first place.

Existing technologies could not have prevented the emergence of the novel coronavirus. However, AI systems could have better tracked and could have even predicted new outbreaks. Technology could have helped providers treat more patients with fewer hands. Having a robust virtual care system in place would have been preferable to rolling it out ad hoc over many months as we did. Better AI analytic systems could have rapidly identified best practices for limiting the spread of disease and providing better treatment for new cases.

In the pandemic's wake, we are now seeing these technologies implemented at greater scale throughout the industry. Many healthcare organizations turned to underutilized technologies to address the crisis. The coronavirus presented challenges unprecedented in modern history. Healthcare organizations tried to address them with digital solutions that could be developed and implemented quickly. This triggered a phase of rapid experimentation and innovation. This was certainly true for Anthem. In the first three months of the pandemic, we launched several new digital programs aimed at addressing the crisis. Many of these solutions provided benefits in areas outside of the coronavirus pandemic. We have since applied solutions originally developed in response to the pandemic to substance abuse services, cardiac rehab, extending care to underserved patient populations, and a host of other areas.

This rapid phase of experimentation during the pandemic has not been unique to Anthem. Healthcare executives at other organizations report similar experiences. The pandemic drove innovation across the entire industry as organizations looked for digital solutions to unprecedented challenges. The healthcare system was faced with a black swan event that forced it to deal with its deficiencies—in particular, its failure to adopt new technologies rapidly. These advances will remain even after the pandemic ends. COVID-19 has forever changed the way we consume, deliver, and pay for healthcare. The pandemic has—paradoxically—shored up and strengthened the healthcare system. We are now better prepared for the next pandemic or other crisis as well as for routine operation.

Unfortunately, the price paid for this period of advancement has been enormous. A half million Americans and counting are dead and untold more are struggling with ongoing complications. Healthcare workers are dealing with PTSD and burnout. Many have or will leave the profession forever. The economic toll from shutting down the economy to control mass outbreaks has cost us trillions of dollars and millions of lost jobs.

Much of this pain and suffering could have been avoided if we had been more prepared in the first place. A proactive response beats a reactive one—facing a crisis prepared beats both. Organizations that already had up-to-date digital solutions in place fared best. Those that were caught off guard but that reacted with the future in mind did second best. They didn't simply react to the crisis at hand. They recognized the deficiencies that the crisis exposed and updated and strengthened their systems permanently.

In this way, the pandemic has permanently accelerated the technology adoption curve across the entire industry. Going forward, healthcare organizations that lag behind in adopting new technologies are going to be less viable than before the coronavirus. Organizations that don't embrace these digital solutions will fall behind and suffer. The healthcare industry is finally starting to catch up with technologies already being applied in other industries. It only took a global pandemic, a half million dead, tens of millions of lost jobs, and a year of hardship and civil strife to get us there.

None of this was a forgone conclusion. Catastrophes happen—but catastrophes that could have been prevented or better mitigated by existing technology are an *unnecessary* risk. Black swan events like the coronavirus pandemic can jolt us into action and accelerate the adoption curve, but they are not the ideal mechanism for doing so. Ideally, we would implement new technologies before a crisis forces us to play catch-up. Reactive responses to crises are almost always less effective and cost efficient than implementing best practices and optimal solutions before disaster strikes. However, this requires proactive experimentation in digital solutions and new technologies so that we don't fall behind in the first place.

Accelerating the Adoption Curve through Active Experimentation

Deploying optimal digital solutions isn't as simple as flipping a switch. Organizations need a vision and a plan. They need to consider their particular risks and vulnerabilities as well as any regulatory complexities. Missteps can be expensive. Substantial up-front layouts must be made. Legacy IT systems often need to be overhauled or replaced. Staff need training on new systems and procedures. None of this is cheap. Getting things wrong can mean millions of dollars wasted only to have to roll systems back. The risk is substantial, which is why healthcare organizations are slow to adopt new technologies in the absence of industry pressure or a sudden crisis.

This is a mistake. Novel technologies don't merely offer solutions. They help us identify problems, such as the deficiencies exposed by COVID-19. These problems are often specific to particular systems. Different organizations have certain needs and face different challenges. There are no one-size-fits-all solutions. Organizations must find their own personal optimal solutions. Digital solutions allow for precision-targeted customization, but this means that the optimal solution will be specific to the organization and the problem at hand. Thankfully, AI analytics can unlock insights in your data that help identify *your* problems so that you can find *your* optimal solutions. Without AI, we couldn't parse the massive amounts of data generated today. AI-enabled data analysis teaches us how to best employ new technologies. These are the most powerful tools at our disposal. We absolutely must be using the best tools if we hope to find the best solutions.

You also need the best data. Insights that lead to optimal solutions can be extracted from relevant data. Organizations are not always collecting or even generating all of the data they need. There is a difference between the data you *have* and the data you *need*. My job at Anthem for the last three years has been to spearhead and oversee innovation and experimentation with digital technologies. In trying to transform Anthem into a digital-first company, we quickly realized that our success depended on sound technological experimentation in business-relevant applications. Experimentation taught us what we didn't know—the "unknown unknowns." Active experimentation helped us better understand our organization's workflow and business environment, and it produced the data we needed to find optimal solutions. The data you have gets you headed in the right direction. The data you produce through experimentation makes it possible to correct course and fine-tune. That's the only way to get to *optimal* solutions. This is not a buzzword—by *optimal* I mean the best possible solution to a problem using all available tools at your disposal.

Active experimentation in the application of technology to relevant business problems has been the key to Anthem's transformation into a digital-first company. We engage in safe, business-relevant experimentation in order to understand our deficiencies and identify the best solutions. We are a company transformed by this methodical, targeted experimentation. Now we want to become part of an *industry* transformed. Novel exponential technologies have radically transformed other industries, from retail to FinTech. Healthcare has been insulated from the same competitive pressures as those industries. This has been to the benefit of companies lagging behind on technological adoption but to the detriment of the industry and society.

Although it pains me to say this, Anthem isn't special, not here. Other companies can force the same changes. The industry doesn't have to wait for the next black swan event to press the adoption of new technologies and digital solutions already deployed in other industries. We can also simply choose to be innovators. With enough willful innovators in the healthcare space, the laggards will be forced to keep up. In this way, those of us in leadership have the power to accelerate the technology adoption curve across the entire industry.

To that end, Anthem has developed a model for building out rapid, reliable experimentation in digital solutions. The Digital Data Sandbox is a data set and analytic system for organizing and analyzing deidentified healthcare data (both internal and external), which researchers and organizations can use to test their own models and to search for their own optimal solutions. The Digital Data Sandbox is both the product of and a tool for active experimentation with healthcare data. Our partnering with other organizations helps them use our model to find winning solutions while growing our platform's data set. The Digital Data Sandbox is now the largest single-payer certified deidentified data set in private US healthcare. In the first six months of 2020, researchers and data science experts logged over a million minutes of algorithm and model development time on our platform. These users have been mining twelve billion claims across fourteen years of data. The system now has over two thousand data fields on more than forty million patients and consumers.

These data sets are made publicly available online. The platform has been used for multiple hackathons and global challenges as well as for healthcare research and in practical business applications. We have evaluated over two hundred use-case opportunities and launched multiple digital solutions to real-world problems. This is in addition to our own private use of the platform at Anthem, where the Digital Data

Sandbox is used to find better ways of handling business processes and designing technology systems.

We are rapidly identifying optimal solutions to our specific problems and needs and helping others do the same. The result is a company transformed. The destination is an industry transformed. We're creating innovators with the goal of forcing the competition to keep up. Even without black swan events, the opportunity cost of ignoring novel technologies is too great. We aim to force the industry's hand.

The False Dilemma of Making Sensitive Data Shareable

This kind of digital experimentation relies on data sharing for analytics. The best solutions are derived from the best available data. Incomplete (or bad) data leads to suboptimal or even flawed outcomes. Both internal business data and external industry data are necessary. Unfortunately, fear of data breaches has caused many organizations to be risk averse when it comes to sharing data. This is especially true in the healthcare industry, where patient data is particularly sensitive. Healthcare companies can face serious legal repercussions in the event of a data breach, even a relatively small one. Many healthcare executives have concluded that making data more shareable is often not worth the risk.

These fears aren't without merit. High-profile data breaches keep hitting the headlines. The healthcare industry has not been unaffected, and the problem has only been getting worse. Data breaches and phishing attacks compromised five million patient records in 2017, fifteen million in 2018, and over twenty-five million in 2019, according to the Protenus Breach Barometer, which tracks data breaches in the healthcare industry.

Healthcare organizations have responded by keeping data siloed in databases on premises. Access to data is kept extremely limited to make security simpler, safer, and more cost effective. Data that rarely changes hands is less likely to fall into the wrong hands. Unfortunately, these draconian security measures keep out good and bad actors alike. The data is safer from hackers but often inaccessible to scientists and researchers who could put it to use for analytics.

The fragmentation of healthcare data stifles collaboration between organizations and puts a drag on innovation across the whole industry. Finding the optimal solutions to the problems facing the healthcare industry will take the best and brightest minds working with a full arsenal of tools. Data is one of our primary weapons in the fight. In 2020, the healthcare industry generated approximately two thousand exabytes of data. Unfortunately, more than 99 percent of this data is locked away in isolated dark corners of various enterprises, organizations, and health systems where it is often not available for analytics.

We simply cannot allow this data to go so criminally unused. The stakes are too high. Healthcare industry data contains insights that will help prevent or mitigate black swan events looming on the horizon. COVID-19 will not be our last crisis. Every day that we deny ourselves the insights in this data is its own crisis. The innovative and revolutionary technologies

> Organizations can find new and better solutions to business problems by gleaning insights from industry data—but only if the industry makes that data shareable for analytics.

described in this book absolutely rely on the insights locked away in this data.

Organizations can find new and better solutions to business problems by gleaning insights from industry data—but only if the industry makes that data shareable for analytics. Security protocols that limit access to data stifle collaboration and ultimately slow the adoption of new technologies and limit their utility. For example, consider unsupervised machine learning. Data analysis has traditionally been performed on data that has already been cleaned, labeled, and structured. Simple algorithms crunch the data just as they are programmed to do. Unsupervised machine learning is revolutionary because it searches for patterns within *unstructured* data. This is essentially how artificial intelligence identifies the unknown unknowns in data—those insights that data analysts didn't know to look for. This is a powerful tool for finding better digital solutions, but it relies on unfettered access to data. The more data, the better. Machine learning generates optimal insights only when all available data is used. Without full access to data, machine learning technologies are less useful and therefore less likely to even be adopted.

Researchers working in machine learning often have the hardest time accessing healthcare data. Security protocols frequently require researchers to show how their experimentation will provide a benefit. But researchers working in machine learning are looking for those unknown unknowns that can be found only by working with the data. This leaves them in a catch-22. They have to show why they need access to specific sets of data in order to gain access. However, they cannot prove that need without first exploring the data. You cannot report on the benefits of insights you haven't generated yet. Machine learning researchers have to work with data to show what valuable insights it holds.

This is obviously impossible, so these unknown unknowns often remain unknown and their value uncaptured. Many healthcare organizations simply don't care. They place a premium on privacy and security and are unwilling to make data more shareable given the perceived risks of doing so. Keeping data siloed and fragmented across the industry seems like a small price to pay for better ensuring the security and privacy of data.

> We no longer have to forgo privacy or security to share data for analytics. The tension between data utilization and privacy/security has become a false dilemma.

This might be a reasonable decision *if* it were a choice we actually had to make. However, this is no longer the case. Advancements in cryptography, blockchain, cloud computing, and machine learning now allow for the safe sharing of *deidentified* data. Data can be shared and accessed without compromising privacy or security. These technologies allow us to experiment with sensitive data safely and securely. We no longer have to forgo privacy or security to share data for analytics. The tension between data utilization and privacy/security has become a false dilemma.

None of this is to say that privacy or security are unimportant or afterthoughts. On the contrary, they are absolutely foundational. Privacy and security are nonnegotiable issues when it comes to data, especially healthcare data. Data systems and digital solutions need to be designed and managed so that data is kept safe. Privacy and security are the *first* considerations when designing digital solutions. Plans can be altered, protocols changed, algorithms edited, and digital solutions modified. However, you cannot undo a data breach. Once data is leaked, it's leaked. There's no getting the cat back into the bag.

Data safety, privacy, and security are therefore the absolute first things we consider when working with sensitive data. Privacy and security cannot be compromised in any way, but with blockchain and other cryptographic advancements, their nonnegotiable nature doesn't have to preclude the sharing and use of data.

Prototyping—Start Small, but Plan for Scale

Privacy and security are the primary nonnegotiable needs. Once plans for keeping sensitive data safe are in place, organizations can begin rapid experimentation with new digital solutions. Prototypes should be limited in scope so that they can be deployed and tested quickly. Putting your prototype to use generates data that can help you refine it later. Initially, you want to limit features to those that are necessary for a proof of concept. Experimentation is an iterative process. Features can be added and expanded upon later as long as you plan with future growth in mind. Planning out optimal solutions entirely ahead of time is not only inefficient but also impossible. Experimenting with technologies in business-relevant use cases is the only way to generate the data that will lead you to *optimal* solutions.

This is the process we employed when building the Digital Data Sandbox. We debuted our prototype in November 2018, only four months into my tenure with Anthem. The purpose of the project was to build a system for safely and securely sharing healthcare data for analytics. The first step was, of course, to ensure the safety and privacy of the data. We built the platform around the sharing of *deidentified* data, which could be transferred securely without posing a privacy risk, even in the event of a data breach. We then built a simple prototype. Our ambitions were grand, but our first attempt was modest. The original prototype included only a few dozen data fields. Functionality was intentionally limited in scope, but the prototype was sufficient for researchers, entrepreneurs, and other innovators to start experimenting with the system.

Our initial feedback wasn't great. Several start-ups applied to use the system to test various digital analytic solutions. One of our first users was CloudMedx, an innovative start-up known for experimenting with exponential technologies, which wanted to use the Digital Data Sandbox to test an algorithm for predicting hospital readmissions. Their compute was insufficient due to sparsely populated data fields. The system also suffered from too much latency. The experience was generally poor.

Undeterred, we used feedback to refine the prototype. In May 2019, we released an updated version with added features and an expanded data set that included data on forty-five million unique people and seven billion claims from across twelve years. The new model proved to be much more useful and user friendly. New users tested a variety of use cases on our system, including models for predicting the development of comorbidities and algorithms for determining best practices to prevent cancer recurrence. This prototype worked much better and allowed for a phase of rapid experimentation with the model. We ran two hackathons on the data set, began research with academic institutions, evaluated more than one hundred third-party use cases, and tested over twenty of them.

In July 2020, we released a third version of the Digital Data Sandbox with an additional five billion claims and two more years of data. We also restructured the data to include more connections and associations. The model was now streamlined and well suited to rapid experimentation. The system was much more usable so that it could be deployed in far more use cases. These experimentations generated their own data and insights that further enlarged and refined the data set. Experimentation increased our capacity for additional experimentation exponentially. The better the model became, the faster we could conduct targeted experiments to improve it further.

We created the Digital Data Sandbox with this iterative process in mind and helped researchers and organizations fine-tune their own digital solutions on our platform in the exact same way. Healthcare organizations now have a new tool for building, testing, and using scalable prototypes on the largest set of sharable, secure, deidentified healthcare data in existence. We built a digital solution for building digital solutions. Nonnegotiable privacy and security controls are built into our model and we help partners build them into theirs. They are able to use our model to fine-tune their own prototypes. We encourage them to include end decision makers early in the process so that they, too, design with future growth in mind. This allows their models to scale up just as quickly as ours. Change management features can be built right into their prototypes, just as they are embedded in the Digital Data Sandbox.

The Digital Data Sandbox is a testament to our ethos that the best way to find optimal digital solutions is rapid experimentation with business-relevant use cases. We found an optimal solution for developing optimal solutions within the healthcare space. This sort of layered iterative experimentation is helping us build a better and brighter future for healthcare.

The Future of Healthcare Is Digital ... with a Human Touch

These kinds of digital solutions are not going to remain optional. Technology adoption is finally picking up in healthcare, just as it has in other industries. More innovators are entering the healthcare space. More legacy organizations are transforming themselves into innovators. We have had our black swan event in COVID-19. There will be others in the future. But change is already in motion. There's no going back. Going forward, the price organizations pay for lagging behind on adopting new technologies and developing better digital solutions will be much higher.

Digital will soon be the default mechanism of care. Within three to five years, as much as 80 percent of nonacute specialty care will be delivered virtually. This includes primary care as well as women's health, transgender care, elder care, pediatric care, and more. If you're not coming in on an ambulance or a helicopter, the likelihood that much of your care will be virtual is high. Virtual care will become the norm wherever possible.

As virtual care becomes increasingly common, consumers will start expecting the same level of quality as they receive in person. They will demand digital healthcare services that are as efficient and user friendly as those in other markets. Consumers become rapidly accustomed to seamless digital consumer experiences. They will soon expect online healthcare to be as seamless and efficient as online shopping. Healthcare organizations are going to have to deliver them to remain competitive.

However, virtual care will not replace traditional care. Virtual and in-person care will instead become increasingly interchangeable and integrated as digital solutions are applied throughout the industry. Healthcare consumers will expect digital solutions that make going in for care as easy, efficient, and effective as possible. This will mean user-friendly patient portals that provide a customized experience. It will mean care protocols that have been studied through analytics to provide the best experience possible. It will mean the most effective custom-tailored therapies delivered efficiently. Winning digital solutions will be those that use the most innovative technologies available to deliver a superior and seamless experience.

Behind the scenes, more and more of the healthcare industry will be powered by exponential technologies. New digital solutions are spreading across the industry and transforming the way we provide care and do business. Hospitals will expect optimal digital solutions from their vendors, suppliers, and partners. Care teams are going to demand workplace systems that leverage technology to make their jobs easier and their work safer and more effective. The bar is rising across the whole industry. The days when healthcare organizations could coast by without experimenting with new technologies are coming to a close. The industry is finally catching up, and laggards will soon be left behind.

Because the optimal digital solutions vary by use case and context, the winning solutions for individual companies won't be the same. Off-the-shelf digital solutions will probably exist in some cases, but they won't always be optimal. The most successful companies and organizations in a digital healthcare era will be those that don't just use exponential technology but use exponential technology to understand, define, and solve their problems. Machine learning, AI, and other technologies—with the privacy and security protections of blockchain and advanced cryptography—will make this not only possible but also necessary. Organizations will have to experiment to identify unknown problems and deficiencies and to unlock unconsidered solutions.

The problems the healthcare industry faces, as described previously in this book, are substantial but not insurmountable. We know how to build winning solutions. They will feature radical interoperability so that systems can work together synergistically. They will be built around privacy and security from the ground up so that data can be best collected, generated, and actually used for analytics. They will solve problems we don't even know we have—those unknown unknowns—through rapid experimentation on business-relevant use cases. Optimal solutions will be customizable and custom tailored to meet consumer promises and expectations while creating meaningful market differentiation that forces competition. They will leverage the full potential of data and novel technologies to fully understand the problems we face and solve them as best we can with the tools we have. It's going to result in a radical transformation of the American healthcare industry.

As healthcare leaders, we have a duty to oversee this transformation and hurry it along. We have already waited too long. It is up to us to make sure that our own companies and organizations are leading the way. These changes are not to be feared. They should be embraced and actively accelerated. Rapid experimentation is the key to getting the virtuous flywheel of individual choice that we covered in the last chapter turning, faster and faster, as it carries us into the future.

It's up to us, as industry leaders, to get it turning.

Let's make it so.